in the wake of the *Beagle*

SCIENCE *in the* SOUTHERN OCEANS *from the* AGE *of* DARWIN

Edited by IAIN McCALMAN
and NIGEL ERSKINE

UNSW PRESS

AUSTRALIAN NATIONAL MARITIME MUSEUM

AUSTRALIAN
NATIONAL MARITIME
MUSEUM

A UNSW Press book

Published by
University of New South Wales Press Ltd
University of New South Wales
Sydney NSW 2052
AUSTRALIA
www.unswpress.com.au

National Library of Australia
Cataloguing-in-Publication entry
Title: In the wake of the *Beagle*: science in the southern oceans from the
        age of Darwin/editors Iain McCalman, Nigel Erskine.
ISBN: 978 1 921410 94 9 (pbk.)
Subjects: Darwin, Charles, 1809-1882 – Influence.
        Evolution (Biology)
        Science – History
        Navigation – History.
Other Authors/Contributors:
        McCalman, Iain.
        Erskine, Nigel.
Dewey Number: 576.82

Design Di Quick
Front cover iStockPhotos of Blue-footed booby and Blue Morpho butterfly
Back cover Turtleshell and wood mask collected from Naghir
        (Mount Ernest) Island in 1849. (© The Trustees of the British
        Museum)
Printer Everbest, China

contents

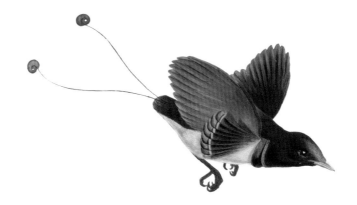

contributors

LISSANT BOLTON is head of the Australian and Pacific section at the British Museum. Her anthropological research focuses on Melanesia, especially Vanuatu. Her publications include *Unfolding the Moon: Enacting Women's Kastom in Vanuatu* (University of Hawaii Press 2003). Her current major research project (2005–2010) reconnects Melanesians with collections from that region held by the British Museum.

JIM ENDERSBY is a historian of science at the University of Sussex. He is the author of *Imperial Nature: Joseph Hooker and the Practices of Victorian Science* (2008).

NIGEL ERSKINE is curator of exploration at the Australian National Maritime Museum. He has a strong maritime background combining academic qualifications in history and maritime archaeology with a trade certificate in ship and boat building. He has undertaken archaeological work on several eighteenth-century British naval vessels lost in the Pacific and completed his doctorate on the historical archaeology of Pitcairn Island in 2004.

JULIAN HOLLAND is a researcher and former museum curator specialising in historic scientific instruments.

SOPHIE JENSEN is a senior curator at the National Museum of Australia. She is currently undertaking her doctorate examining the life of the naturalist John MacGillivray at the Research School of Humanities, Australian National University.

IAIN McCALMAN is a professorial fellow in the Department of History, University of Sydney. In 2007 he was made an Officer of the Order of Australia for service to history and the humanities as teacher, researcher, author and through advocacy, advisory roles in academic and public sector organisations. He is currently working on a book about Darwin, Hooker, Huxley and Wallace, *Darwin's Armada: How Four Voyagers to Australasia Won the Battle for Evolution and Changed the World*.

RICHARD NEVILLE is the Mitchell Librarian, at the State Library of New South Wales. He has published widely on nineteenth-century colonial art and culture.

JUDE PHILP graduated from Darwin College, Cambridge for her doctorate on Torres Strait Islanders' history. Now senior curator at the Macleay Museum, University of Sydney, she has published on photography, ethnography and nineteenth-century European and Pacific Islander encounters.

KATE WILSON is director of the CSIRO's Wealth from Oceans National Research Flagship, which seeks to increase social, environmental and economic wealth from Australia's marine territory, and to understand the complex relationship between ocean, land and atmosphere, particularly in relation to climate change. Her research expertise includes molecular biology and biotechnology in agriculture and aquaculture. She has consulted to the International Atomic Energy Agency, co-founded CAMBIA and led tropical aqua-culture research at the Australian Institute of Marine Science.

foreword

Mary-Louise Williams
Director, Australian National Maritime Museum

This year marks both 200 years since the birth of Charles Darwin, and the 150th anniversary of the publication of his most famous work: *On the Origin of Species*. To celebrate this significant occasion, the Australian National Maritime Museum has worked in close collaboration with a number of institutions to create the exhibition *Charles Darwin – Voyages and Ideas that Shook the World*. The exhibition brings to Australia for the first time objects directly associated with Darwin's voyage aboard the *Beagle*. These include Conrad Martens' stunning South American watercolours made while voyage artist, Captain Robert FitzRoy's 1836 survey of the Cocos Keeling Islands, some of Darwin's collection of crabs, and objects belonging to John Lort Stokes. Added to these international loans is a rich vein of material from local collections highlighting some important Australian connections. One of these revolves around the role of Phillip Parker King who as expedition leader commanded the *Adventure* and *Beagle* during the first South American survey.

King was a talented artist and it is wonderful to see his work placed in context with the *Beagle's* later voyages.

However, the exhibition is not the Maritime Museum's only activity celebrating Charles Darwin. As a research partner in the Australian Research Council Linkage Grant *Seeing Change: Science, Culture and Technology from the Age of Darwin*, the museum worked closely with Professor Iain McCalman of Sydney University and Kim McKenzie of the Australian National University in organising the Darwin symposium held at the Museum in March this year. Charles Darwin is associated with an enormous variety of stories and the symposium brought together speakers from all over the world, providing the opportunity to delve deeper into the stories raised in the exhibition. The symposium also provided an opportunity to capture this research and to make it available to a wider audience through a publication. This book is the result.

This is an important work and it could not have been

acknowledgments

completed without the assistance of a number of institutions. I wish to acknowledge the assistance of the British Museum, the National Maritime Museum (UK), Oxford University Museum of Natural History, the State Library of New South Wales, the Macleay Museum, the Powerhouse Museum, the National Library of Australia, the Tasmanian Museum and Art Gallery, the State Library of Tasmania, CSIRO and the Australian Research Council.

Charles Darwin spent only three months in Australia but he shared broad links with this country and his work continues to inspire scientific projects in modern Australia today.

I would like to thank the following people for their assistance with this project:

The *Charles Darwin – Voyages and Ideas that Shook the World* exhibition team: Susan Bridie, Nigel Erskine, Jeff Fletcher, Richella King, Michelle Linder, Will Mather, Daniel Ormella, Gemma Nardone, Johanna Nettleton, Lindsey Shaw, Daniel Weisz and Caroline Whitley.

Richard Neville, Jim Endersby, Sophie Jensen, Jude Philp, Lissant Bolton, Julian Holland, Kate Wilson. Sammy De Grave, Margot Riley, Tony Marshall, Marco Duretto, Kim McKenzie, Kate Bowan and Meg Rive.

Special thanks to Professor Iain McCalman for a very successful collaboration and to Katherine Anderson for locating and managing all the illustrations and for co-ordinating the publication.

Michael Crayford, Assistant Director, Collections and Exhibitions, Australian National Maritime Museum

introduction

Nigel Erskine and Iain McCalman

Britannia needs no bulwarks,
No towers along the steep;
Her march is o'er the mountain-waves,
Her home is on the deep.[1]

For a small ten-gun brig belonging to what sailors wryly called the 'coffin class', HMS *Beagle* has created the longest wake of any ship in history. Her fame derives from the passenger whom the captain, Robert FitzRoy, employed as a naturalist-companion on a five-year voyage among the southern lands and oceans from 1831 to 1836. Charles Darwin was 22 when he embarked. He'd been born on 12 February 1809 into a world preoccupied by the changing fortunes of the Napoleonic wars. Just four years after Nelson's victory at Trafalgar, Napoleon's armies were firmly entrenched in Spain and the young general Arthur Wellesley was yet to lead a grindingly slow campaign across the Iberian peninsula and on to final victory in France in 1815. Waterloo marked the end of more than twenty years of almost continuous warfare across

Europe and when the fighting finally stopped Britain found herself the undisputed superpower of the world.

England had the world at her feet. She was the first industrial nation, and blessed as she was with apparently limitless supplies of coal and iron, during the past fifty years she had so mastered the mechanical arts that she had outstripped all her competitors. The British stood at the threshold of a colossal boom, for they possessed a virtual monopoly of the techniques of steam, which was presently to prove itself the basic energy of the age.[2]

The world in which Charles Darwin received his early education and training was thus experiencing rapid and fundamental industrial change. As it happens, he and his family were among the beneficiaries. His uncle, Josiah Wedgwood, had instigated new technology, transport and marketing innovations to become the country's foremost pottery tycoon, and Charles's father, Dr Robert Waring Darwin, had made himself independently wealthy by judicious investment in the boom for railway and canal building. Meantime, the monarchy of George III sputtered to an end, to be replaced by that of his unpopular eldest son, first as Prince Regent and then for a decade in his own right as George IV (1820–1830). During the same period, Britain was increasingly wracked by mass radical demonstrations, fomented by the middle and working classes, for political and economic reform in a country still formally governed by an aristocratic elite. This period also saw unprecedented demographic changes, as the rural population of England was drawn to work opportunities in the new industrial centres of Manchester, Liverpool and Birmingham. By the time that Darwin set off for South America on board the *Beagle* in 1831, London had become the largest city in the world – a status it would continue to hold for almost 100 years.

The post-Napoleonic peace also heralded a new era of opportunity to consolidate on the late eighteenth-century voyages of discovery and to continue exploration beyond the established sea routes, to penetrate the polar regions, and to push beyond the existing fringes of European settlement. This new phase of exploration was in large part the brainchild of John Barrow, Second Secretary of the Admiralty Board, a one-time explorer himself, as well as a wily civil servant with indefatigable powers of persuasion. An incorrigible geographic 'projector' who wrote regularly for the influential *Quarterly Review*, he urged the Admiralty to deploy Britain's demobilised naval officers – at the time unemployed on half-pay – as a peace-time force for extending Britain's imperial power. They would use their naval expertise to 'discover' unexplored lands, to map and chart sea passages and to open up trade routes and imperial settlements in New World regions like South America, Africa, Australia and the Arctic and Antarctic circles.[3]

A key ally in this enterprise was Captain Francis Beaufort, the newly appointed Royal Hydrographer, a naval veteran who believed that maritime sciences of hydrography, magnetic and astronomical charting, and weather prediction could revolutionise the navy and in the process extend Britain's geopolitical and economic reach. He hoped that the better-trained young captains, skilled in naval sciences, would produce a new kind of surveying that would make use of the latest scientific knowledge and precision technologies. He aimed to generate a scientific girdle for Britannia, a chain of accurate measurements that would circle the world.[4] HMS *Beagle*, the ship that carried Charles Darwin on his famous voyage, was employed in just such service, sailing more or less continuously for 17 years in the course of three successive commissions (1826–1843).

As the nineteenth century unfolded, science and technology combined to reveal the first coherent picture of the world, based on an unprecedented array of evidence gleaned from explorers, collectors and a new breed of specialist scientists bringing together information from fields such as geology, anatomy, paleontology and botany, and all leading to a radical reappraisal of the earth's age and the origins of life.

A year after Darwin returned to England with the *Beagle*, the 18-year-old Princess Victoria was crowned queen, commencing a 63 year reign which would see Britain establish the largest empire on earth and her colonies grow to maturity and self-government. In Australia, the nineteenth century witnessed the transformation of British power from a single coastal settlement exercising relatively tenuous control of a small part of the east coast in 1800, to the establishment of largely autonomous colonies occupying the entire continent and populated by almost four million Europeans by 1900.

The nineteenth century was also supremely a period of change during which the technologies and machines of a commercial and industrial age were harnessed to create a spectacular new world, designed not by God, but by Man. The poet William Wordsworth, crossing Westminster Bridge, was forced to pause in awe of the view of London before him and to capture the moment:

> Earth has not anything to show more fair:
> Dull would he be of soul who could pass by
> A sight so touching in its majesty.[5]

In the middle of the century, the national pride in British innovation and imperial advance was celebrated most publicly in the glass and iron novelty of the Crystal Palace's Great Exhibition. Visited by six million people during its five-and-a-half month season in 1851, the exhibition was an international wonder, where enormous iron machines, delicate Wedgwood china and works of art were viewed with equal rapture.

The Great Exhibition was an unprecedented success, illustrative of the confidence of a nation at the forefront of industrialisation and demonstrating the progress made over the past 50 years. Less universally celebrated – but of far more lasting significance – Charles Darwin's *On the Origin of Species* was published eight years after the Great Exhibition and demonstrated a parallel intellectual progress of scientific understanding of the world.

As Darwin fully acknowledged in the introduction to his book, his ideas were a legacy of the work of others and formed but a step in 'the progress of opinion on the Origin of Species'.[6] Evidence of the evolution of species had been realised for some time and had been steadily gaining recognition throughout the nineteenth century. The work of the Comte de Buffon, Jean-Baptiste Lamarck, Geoffroy Saint-Hilaire, Thomas Malthus, Alfred Wallace, Thomas Huxley and Joseph Hooker all contributed to the formation of Darwin's evolutionary thinking. What he significantly added was the idea of 'natural selection' to explain the mechanism of change in species.

Darwin might very well have also included to this list of influences the work of his own grandfather Erasmus Darwin and, even more particularly, of the English geologist Charles Lyell. Lyell's *Principles of Geology* – the first volume of which was presented to Darwin by Captain Fitz-Roy on board the *Beagle* – provided both the inspiration for Darwin's study of the formation of coral atolls and a timescale for understanding evolution. Published between 1830 and 1833, in *Principles of Geology* Lyell assembled an overwhelming amount of data, much of it based on his observations around Mount Etna in Sicily, to support the 'uniformitarian' assumption that geological changes had occurred in the past as a result of the same geological agencies acting over long periods at approximately the same rate as they do at present. Contrary to 'catastrosphism', Lyell's work indicated the past was a long and geologically uninterrupted period in which other events could have happened, and it quickly won support. This then provided a setting for scholars to consider the possibility of biological evolution.[7] Another strength of Lyell's work lay in his scientific rigour, and the *Principles of Geology* provided a model which Darwin was to follow in constructing the arguments put forward in *On the Origin of Species*.

One hundred and fifty years after the publication of *On the Origin of Species* it is sometimes difficult to understand the world, and the public attitudes to the question of life, in which Darwin and other scientists worked. The papers that follow address this problem and provide a context for the publication of Darwin's revolutionary theory of evolution by natural selection.

This book presents a maritime view of the voyage of science in Australasia in the period of rapid European expansion during the nineteenth century. Charles Darwin spent only three months in Australia, but as the authors of this work demonstrate, Australasia and the Pacific contributed to the weight of evidence supporting his evolutionary thinking in a variety of ways. Drawn from a broad spectrum of academic and institutional backgrounds, the authors provide new insights into the world of collecting, surveying and cross-cultural exchange in the age of Darwin, as well as glimpses at the 'progress of opinion' in our own era.

In the first part, 'An equal wide survey', Nigel Erskine (Australian National Maritime Museum) and Richard Neville (State Library of New South Wales), respectively provide sweeping panoramas of British naval surveying and shipboard artistic perspectives during our period, setting the scene and capturing the visual images of a long voyaging tradition.

Part two, 'Science on land and sea', offers a series of case-studies of naturalists at work – collecting, observing, preserving, dissecting and analysing the flora and fauna of the southern oceans, islands and continents during British naval survey voyages. Iain McCalman (University of Sydney) looks at the ship-borne trio of Charles Darwin, Joseph Hooker and Thomas Huxley; Jim Endersby (Sussex University) explores the complex reciprocal relationship between Kew botanist Joseph Hooker and his network of colonial collectors and local experts, particularly in Tasmania and New Zealand; and Sophie Jensen (Australian National University and National Museum of Australia) charts the challenges and achievements of one of the period's finest specimen collectors, John MacGillivray.

Part three shifts the angle of vision to 'Indigenous encounters'. Here Lissant Bolton (British Museum) reveals the complex trading relations and daring voyages of peoples encountered by our naturalists in Torres Strait; Jude Philp (Macleay Museum) instills life and meaning into the weapons, tools, ornaments and artifacts collected by the crew of the *Rattlesnake*; and Iain McCalman describes naturalist Alfred Russel

Wallace's dependence on indigenous voyagers, hunters and collectors in making his major bio-geographical and evolutionary discoveries in the Malay Archipelago.

Finally, part four, 'Measuring and mapping new worlds', focuses on maritime science technologies of both the nineteenth century and today, which enable naturalists to describe, understand and utilise new forms of natural knowledge. Julian Holland illuminates the types of instruments that enabled Darwin and his colleagues to bring a new scientific reach and precision to their work; and Kate Wilson (CSIRO) shows how an exciting series of parallel inventions and networks is opening up our understanding of the maritime sciences of today, in an age where global warming has raised the stakes of such work to an unprecedented level.

Strange as it may seem, the long wake of the tiny *Beagle* stretches from the nineteenth century into the future of our globe.

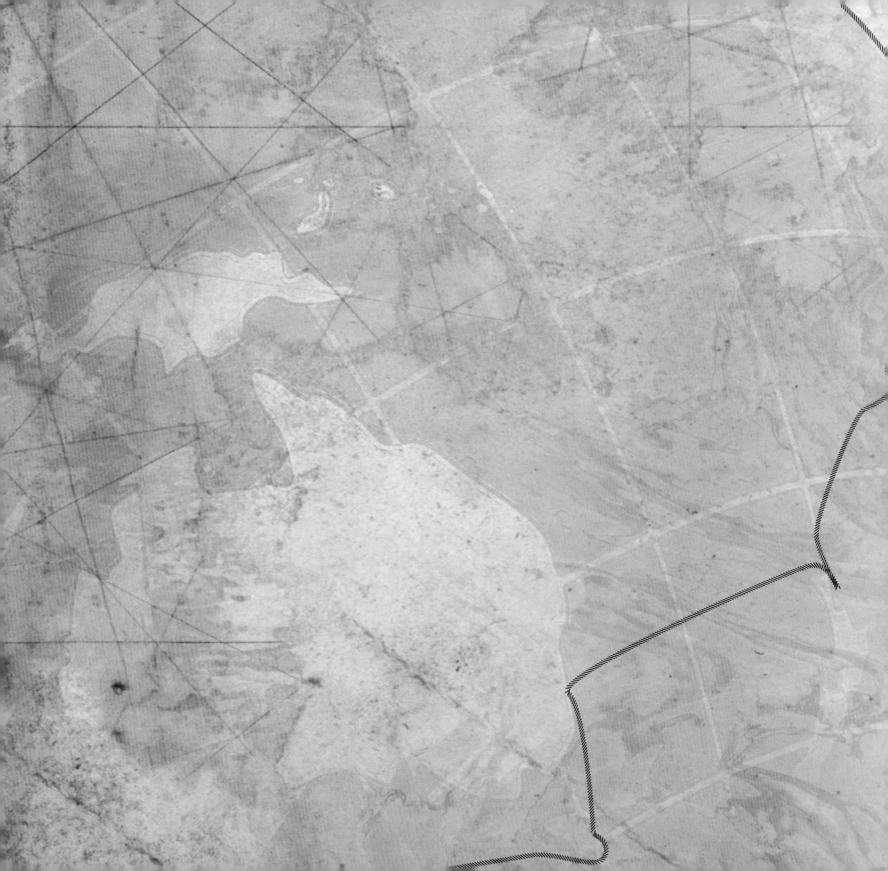

'an equal wide survey'

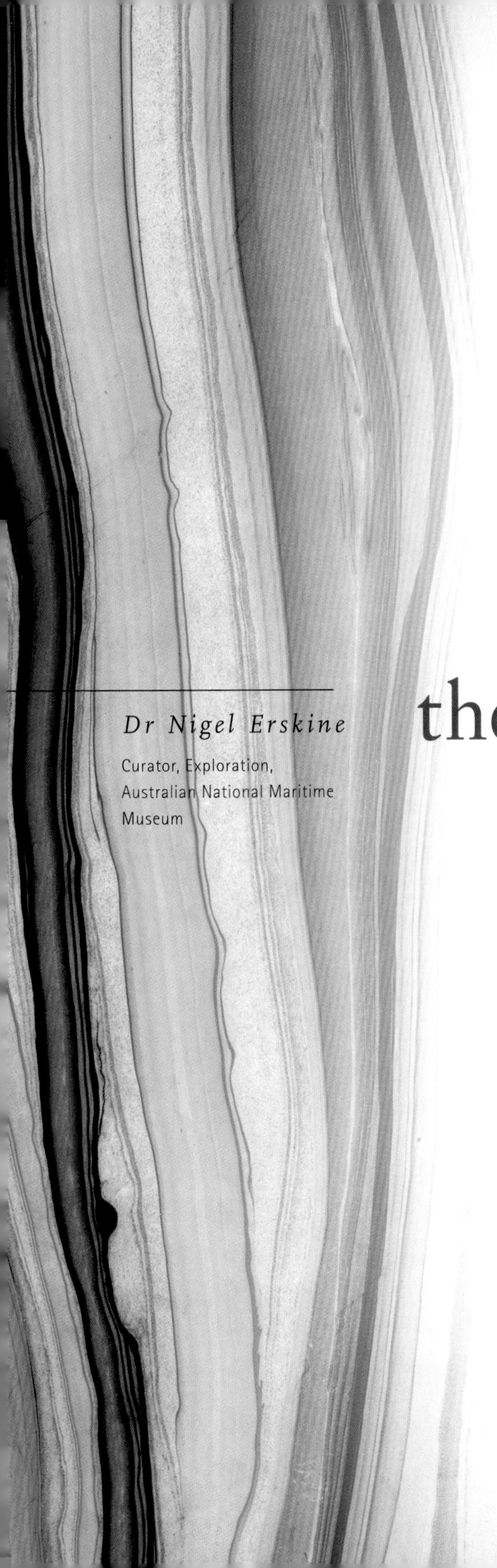

Dr Nigel Erskine

Curator, Exploration,
Australian National Maritime
Museum

the evolution of a tradition

CHARLES DARWIN AND THE AUSTRALIAN WEB

My dear Father. I am afraid I am going to make you again very
uncomfortable. But upon consideration, I think you will excuse
me once again stating my opinions on the offer of the Voyage.[1]

One hundred and fifty years after publication of *On the Origin of Species*, and
at a time when Charles Darwin is recognised as one of the most important scientific thinkers of the modern world, it is interesting to speculate how differently things might have turned out had Darwin's father not been persuaded to
allow Charles to join the *Beagle*. A physician of large practice, ample girth and
good common sense, Dr Robert Darwin raised no less than eight objections to

the proposal. As Charles related to his uncle Josiah Wedgwood (II), his father's objections focused predominantly on the detrimental effects to the young Darwin's career and reputation by being involved in such a 'wild scheme': A two year expedition to South America would be a useless and unsettling undertaking, disreputable to Darwin's prospects as a clergyman, and likely to cause yet another career diversion. And why, if the voyage really presented such opportunity, had no established naturalist accepted the position? Clearly there was a problem – either in the ship itself, or some aspect of the voyage. And in any case, the accommodation would surely be most uncomfortable.[2]

Happily, Dr Darwin's prognosis of the likely outcomes from such a dangerous adventure proved almost entirely correct. Lasting five years, Charles Darwin's voyage aboard the *Beagle* transformed him completely, exposing him to the extraordinary nature of the world and its variety of environments, and planting in his mind the genesis of ideas which would explain the very evolution of life on earth. In later life Darwin acknowledged the seminal importance of the *Beagle*'s voyage from 1831 to 1836 in the development of his thinking. However, such naturalist voyages were not without precedent. Darwin's odyssey was part of a longer process involving earlier ships carrying surveyors and naturalists and would in its turn be followed by other voyages. It was one strand in a continuously expanding web of information and investigation spiralling back to Cook's circumnavigation of the globe in the *Endeavour*.

Joseph Banks and his influence

Independently wealthy after his father's death in 1761, the 25-year-old Joseph Banks came to James Cook's *Endeavour* voyage as a result of his experiences collecting in Newfoundland and Labrador aboard HMS *Niger* in the summer of 1766. This short introduction to voyaging was followed by his election to the Royal Society, where he first heard of the idea to send an expedition to the south seas to observe the 1769 transit of Venus. For Banks the opportunity for expanding his collecting horizons was irresistible, and he focused his money and influence on turning the voyage into a grand scientific adventure – with stunning results.

When the *Endeavour* returned to England in 1771, it brought back around 1400 new plant specimens, in addition to the astronomical data recorded at Point Venus in Tahiti, as well as paintings, native vocabularies and objects collected during the voyage. Cook had, of course, also charted Tahiti, New Zealand, most of the east coast of New Holland and sailed through Torres Strait – discoveries which the Admiralty valued highly. The voyage was undoubtedly a triumph for Cook, but it was, above all others, Joseph Banks who received the public accolades. Seven years later he was elected President of the Royal Society and for the next 42 years he remained the most influential figure at the heart of the British scientific world. To quote John Cawte Beaglehole:

Joseph Banks as 'The Great South Sea Caterpillar', 1795. (James Gillray: Australian National Maritime Museum)

He had found his multifarious calling, and he pursued it,
on the whole, with triumphant success. His role was not to
be an original genius, but a sort of director of scientific and
industrial research.[3]

Lampooned as a glittering butterfly emerging from the south
seas towards the warmth of the royal sun, Banks combined
the power to promote Pacific exploration with his passion for
natural history, facilitating connections between the Admiralty, academics, gardeners and collectors.

One example of Banks's influence is illustrated by Matthew
Flinders' *Investigator* voyage. Flinders had earlier served as a
midshipman aboard the *Providence* on William Bligh's second
breadfruit voyage, and in 1795 he came out to the colony of
New South Wales aboard the *Reliance* carrying the new governor, John Hunter. The establishment and development of
the young colony was an area in which Joseph Banks was
intimately involved and after some notable successes surveying parts of the south coast, Bass Strait and Van Diemen's
Land, Flinders returned to England to seek Banks's support
in mounting an expedition to continue the survey of the vast
continent.

As a survivor of the *Endeavour*'s stranding amongst the reefs
of tropical New Holland, and with the recent loss of HMS *Pan-
dora* on the Great Barrier Reef underlining the dangers, Banks
fully appreciated the importance of establishing safe routes
for shipping to the burgeoning colony. A lull in the war with
France provided a window of opportunity. With growing reali-
sation that the French intended sending an expedition to New
Holland under Nicolas Baudin, Banks had little trouble per-
suading the Admiralty to give Flinders a ship.

In 1776 Banks had moved his natural history collection to
his new London home at 32 Soho Square and installed his
Endeavour companion Dr Daniel Solander as his secretary
and librarian. Juggling his responsibilities as Keeper of Natural History at the British Museum, Solander dedicated a large
part of his time assisting in the enormous task of producing Banks's *Florilegium* – a lavish folio of coloured engravings illustrating 743 botanical specimens collected during
the *Endeavour*'s circumnavigation. (Sadly, the work remained
unfinished at the time of Solander's death in 1782, and the
Florilegium was not published in Banks's lifetime.)

When the *Investigator* finally left England in July 1801, the
extent of Banks's involvement was evident in the number of
supernumeraries aboard. In addition to the naturalist Robert
Brown, the vessel carried the veteran Kew gardener Peter Good,
the natural history artist Ferdinand Bauer, landscape artist
William Westall, astronomer John Crosley and miner John
Allen. Flinders' nephew, midshipman John Franklin, was also
aboard – an experience which may have influenced his later
reception of scientific voyagers when he was appointed Lieutenant Governor of Van Diemen's Land in 1837.

The *Investigator* proved a poor choice, deteriorating rapidly
and causing Flinders to break off the survey in northern Australia. He left the ship at Sydney Cove, and with part of his
crew Flinders took passage aboard the *Porpoise*, intending to
return to England to obtain a more suitable vessel to complete
the survey. But Flinders' luck had deserted him and when

Adventure and *Beagle* at anchor in Possession Bay, Magellan Strait,
Christmas Day 1826. (Phillip Parker King, watercolour: Mitchell Library,
State Library of New South Wales)

the *Porpoise* and another ship, the *Cato*, were lost on Wreck Reef in August 1803, a large part of Robert Brown's plant collection went down with them, along with all Westall's finished sketches. The final disastrous blow to the expedition's fortunes came when, having survived shipwreck, organised a rescue and sailed across the Indian Ocean in a small and leaking schooner, Flinders was made a prisoner of war by the French at Mauritius.

Despite these calamities, his expedition produced important results for natural history. Robert Brown and Ferdinand Bauer had not joined the *Porpoise* and after hearing of the wrecking, set about rebuilding their collections – Bauer spending eight months on Norfolk Island, and Brown collecting in Van Diemen's Land and Port Phillip. Combining material from Banks's collection with his own specimens and notes, Brown had sufficient information to become the authority on Australian botany, and he published the first comprehensive account of Australian plants, *Prodomus Florae Novae Hollandiae*, in 1810. Peter Good had died in 1803, but throughout the expedition he had sent seeds to Banks in England and his actions were responsible for introducing over 100 Australian plant species to Kew.

Ironically, Brown and Bauer returned to England safely on the *Investigator*, and in 1810 Brown became Joseph Banks's botanist-librarian.[4] The situation placed him at the centre of the botanical world and when Banks died in 1820, his will effectively gave Brown control over the enormous collection. When in 1827 the collection was transferred to a newly established department of Botany at the British Museum, Brown was appointed Keeper. A pivotal force at the heart of British botany until his death in 1858, he was a source of information to Charles Darwin. In his autobiography, Darwin recalled of Brown that his 'knowledge was extraordinarily great, and much died with him, owing to his excessive fear of ever making a mistake. He poured out his knowledge to me in the most unreserved manner.'[5] The *Investigator* expedition demonstrated the importance of sending naturalists on surveying voyages, setting a pattern which the Admiralty largely followed for the rest of the nineteenth century.

The Hydrographic Office

With the departure of Matthew Flinders, the survey of the Australian coast remained incomplete and it was not until the end of the Napoleonic wars that the Admiralty again focused on the problem.

Set up in 1795 after the outbreak of war, the Hydrographic Office had struggled under Alexander Dalrymple to supply the navy with charts. Many ships had been wrecked as a direct result of inaccurate charts, and in 1808 a younger hydrographer, Thomas Hurd, was appointed to overhaul the office completely. It is a measure of the importance placed in this work that from 1812 to 1817, when the number of men in the navy was slashed from 145 000 to 20 000, Hurd was successful in promoting the establishment of a dedicated corps of maritime surveyors. As he argued the:

undertaking would keep alive the active services of many meritorious officers whose abilities should not be permitted to lie dormant, whilst they can be turned to national benefit, and would also be the means of acquiring a mass of valuable information that could not fail of being highly advantageous to us in any future war.[6]

The person selected to continue Flinders' work in Australia was Phillip Parker King. The son of Philip Gidley King, himself a veteran of the First Fleet and third governor of the colony of New South Wales, Phillip Parker King was born at Norfolk Island in 1791 and educated in England. He entered the navy at the age of 15 and was promoted to lieutenant in 1814. Three years later he was appointed to undertake surveys in northern Australia, and between 1817 and 1822 he completed five surveys before returning to England.

King's Australian surveys were highly regarded by the Admiralty and in 1826 he was given command of HMS *Adventure* and appointed to lead a survey of the South American peninsula. To assist in this task he was given the 10-gun brig, HMS *Beagle*, commanded by Captain Pringle Stokes. It was during this first South American survey (1826 to 1830) that the *Beagle*'s captain committed suicide and was replaced by Lieutenant Robert FitzRoy. Retiring from active service, Phillip Parker King returned to New South Wales in 1832, where he became a prominent figure and a source of information for all naval commanders visiting the colony.[7]

The South American survey was initiated in response to the break up of Spain's empire. The Spanish colonies had taken advantage of the Napoleonic wars to push for independence and the new republics presented great opportunities for British commerce and trade. However, before these could be exploited, new charts of the South American coast were needed. Part of this work focused on developing viable shipping routes around the tip of South America, which required a thorough investigation of the intricate channels of Tierra del Fuego. In 1830, even at the end of four years surveying, much still remained to be completed. After the *Adventure* and *Beagle* returned to England, the *Beagle* was soon being refitted for its second survey expedition – which would carry Charles Darwin to South America and around the world.

The *Beagle*

Darwin's father was right to question the comfort and safety of the vessel. The *Beagle* was one of more than a hundred vessels, built by the navy between 1808 and 1845, known as 10-gun or 'coffin' brigs – a reference to their unenviable reputation for sinking.[8] Early in FitzRoy's first command, he almost lost the *Beagle* when she was struck by a squall and healed over dangerously. Coffin brigs were notoriously 'wet' vessels, known to ship quantities of water over the deck. In refitting the *Beagle* for the second South American survey, FitzRoy remedied this problem substantially by raising the deck, and fitting a forecastle and poop cabin. The *Beagle* had also previously been fitted with a mizzen mast, altering the sail plan and improving her manoeuvrability.

The *Beagle* below Mount Sarmiento, Tierra del Fuego. (Conrad Martens, watercolour: © National Maritime Museum, Greenwich, London)

HMS *Beagle*. (Conrad Martens, pencil: Dixson Library,
State Library of New South Wales)

Darwin's accommodation was in the poop cabin. At about 1.8 metres tall, he would not have been able to stand upright and the space was so small that he had to remove drawers to make room for his feet when sleeping in his hammock. Shelving across the aft bulkhead housed Darwin's library and the chart table dominated the centre of the cabin. Darwin was also prone to seasickness – a probable inducement for his several overland excursions while in South America.

After completing its work in South America, the *Beagle* sailed west across the Pacific, stopping in the Galapagos Islands, Tahiti and New Zealand before arriving in Port Jackson in January 1836. While the vessel lay at anchor in Sydney, Darwin took the opportunity to explore the country, trekking westward over the Blue Mountains to Bathurst and returning via Phillip Parker King's farm at Penrith. He was unimpressed by the experience and after further short stops at Hobart and King George's Sound he left Australia 'without sorrow or regret'.[9] It was a very different story at the Cocos

Keeling Islands. There Darwin had the chance to examine the reefs closely and to consider them in the light of others he had seen in the Pacific. Years later these ideas were to form the basis for one of his most important publications: *The Structure and Distribution of Coral Reefs.*[10]

By the time the *Beagle* returned to England in 1836, Darwin had been away for almost five years and was thoroughly sick of shipboard life and the sea. Despite this, he fully appreciated the opportunities the voyage had provided and in concluding his journal he could not help but recommend the experience to others. Darwin's *Beagle* voyage was over, but the publication of his narrative three years later brought the voyage to life in the imagination of generations of readers and set an example which other young naturalists were to follow. While Darwin never left England again, the voyage had opened his mind to the world, and especially to the southern lands and oceans, which he continued to explore for the rest of his life through an intricate network of correspondents.

The *Beagle* at anchor under Circular Head, Van Diemen's Land, during its second visit to Australia in 1837–43. (Anon, ink: © National Maritime Museum, Greenwich, London)

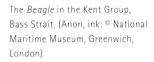

The *Beagle* in the Kent Group, Bass Strait. (Anon, ink: © National Maritime Museum, Greenwich, London)

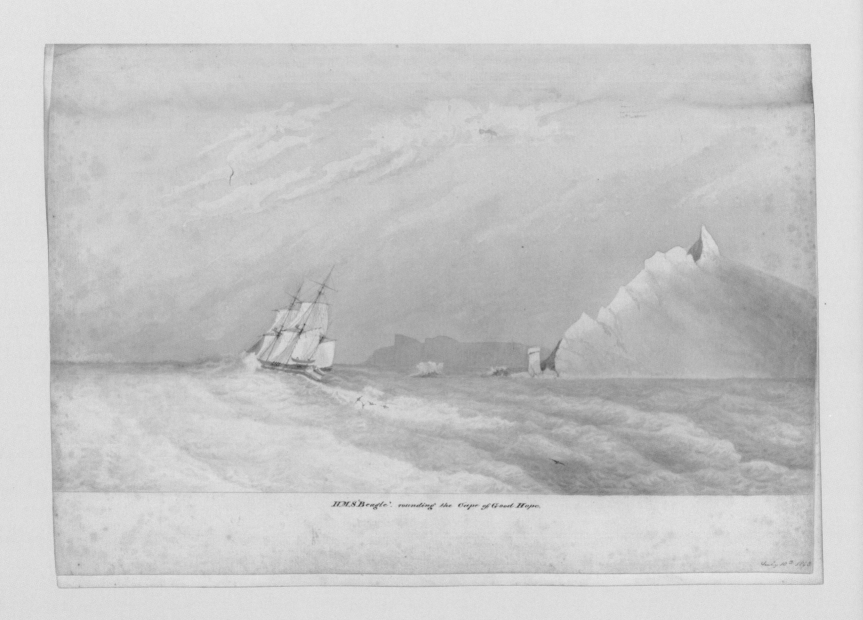

HMS *Beagle*' rounding the Cape of Good Hope.

HMS *Beagle* rounding the Cape of Good Hope at the end of the
1837–43 survey in Australia. (Anon: Lort Stokes Collection,
© National Maritime Museum, Greenwich, London)

The *Beagle* returns to Australia

Whereas his Majesty's surveying vessel *Beagle* under your command, has been fitted out for the purpose of exploring certain parts of the north-west coast of New Holland, and of surveying the best channels in the straits of Bass and Torres, you are hereby required and directed ... to proceed with all convenient expedition.[11]

In early 1837 the *Beagle* was sent back to Australia, where it was to conduct surveys for over six years. Now commanded by FitzRoy's old lieutenant, John Wickham, several other of the crew had also served aboard for both the previous voyages.[12]

The *Beagle*'s third survey voyage was undertaken at a time when Australia's population and commercial activities were expanding rapidly amid growing international interest in the Pacific, Asia and Antarctica, and the commission coincided with the work of several other naval vessels in Australia.[13] Two of these carried men whose support was later crucial in the promotion of Charles Darwin's evolutionary theory: Joseph Hooker and Thomas Huxley.

Hooker and Huxley on the *Erebus*, *Terror* and *Rattlesnake*

Joseph Hooker, the son of botanist William Hooker who was director of the Royal Botanic Gardens at Kew from 1841 to 1865, joined James Clark Ross's expedition to the Antarctic in 1839 as assistant surgeon and botanist aboard the *Erebus*. He collected plants, lichens and seaweed specimens throughout the four-year voyage and later published *The Botany of the Antarctic Voyage of HM Discovery Ships* Erebus *and* Terror *in the Years 1839–1843*. Originally built as bomb vessels – designed to carry mortars and built to withstand their considerable recoil – the *Erebus* and *Terror* were further strengthened and had heating added for the voyage.

During the expedition the ships spent several months in Hobart where they were welcomed enthusiastically by the Lieutenant Governor Sir John Franklin. In Hobart, Hooker was also introduced to local botanists who contributed much to a new authoritative floral compendium of the island, *Flora Tasmaniae*, which was later published as part three of *The Botany of the Antarctic Voyage*.[14]

A life-long friendship later developed between Hooker and Darwin when Darwin invited Hooker to describe the plants in his Galapagos collection. The two shared much in common, especially their fascination with Australian plant distribution, and it was Hooker (along with Charles Lyell) who helped Darwin most in 1858 to establish his claims to

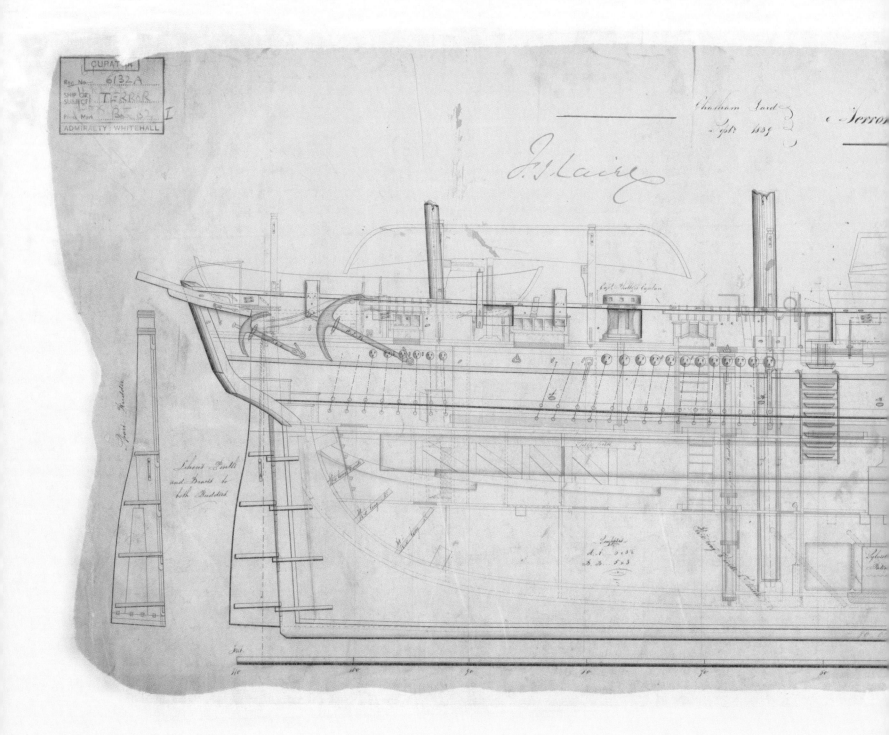

Inboard profile of the discovery ships *Erebus* and *Terror*, after
being converted for polar exploration at Chatham Dockyard in 1839.
(© National Maritime Museum, Greenwich, London)

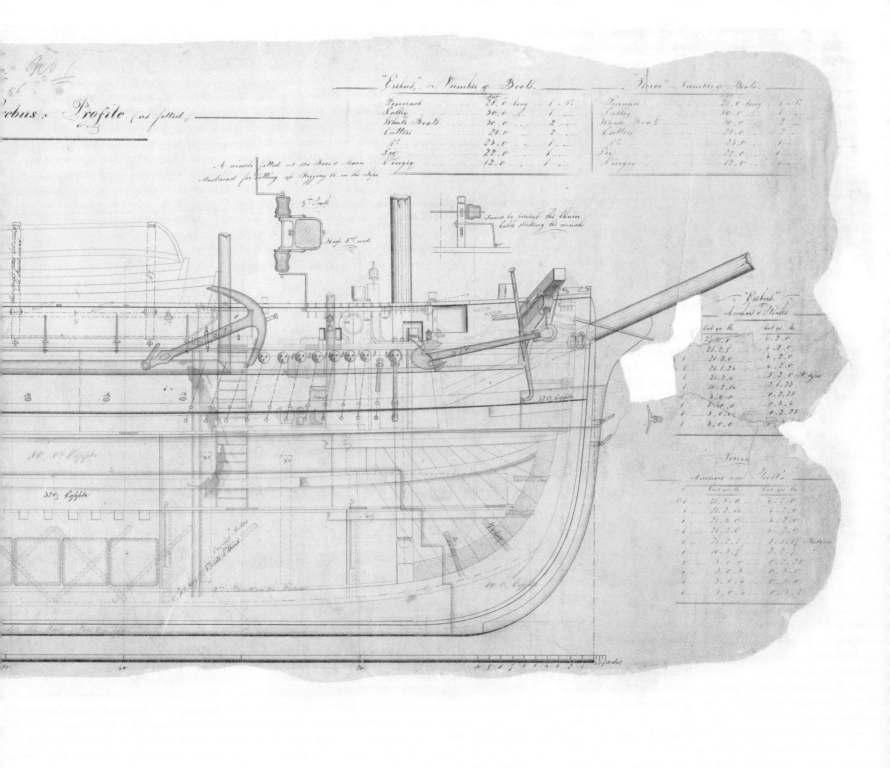

Erebus Profile (as fitted)

"Erebus." Number of Boats.

	feet	
Pinnace	28.0 long	1. N.
Galley	30.0	1
Whale Boats	30.0	2
Cutters	25.0	2
Do	23.0	1
Gig	22.0	1
Dingey	12.0	1

"Terror" Number of Boats.

	feet	
Pinnace	28.0 long	1. N.
Galley	30.0	1
Whale Boats	30.0	2
Cutters	25.0	2
Do	23.0	1
Gig	22.0	1
Dingey	12.0	1

A winch fitted at the Fore & Main Masthead for getting up Rigging &c in the Tops

½" Scale

Hoop 5" wide

Guard to prevent the Chain cable sticking the winch

"Erebus." Anchors & Stocks.

	Cwt qrs lbs	Cwt qrs lbs	
	21.0.0	4.2.0	
	21.2.1	4.2.0	
1	21.3.0	4.2.0	
1	21.1.24	5.2.0	Hedges
1	21.2.4	2.1.23	
1	10.2.14	0.2.28	
1	3.0.0	0.3.4	
1	3.0.0	0.2.23	
1	3.0.0		

"Terror" Anchors and Stocks

	Cwt qrs lbs	Cwt qrs lbs	
N1	21.3.0	4.2.0	
1	21.3.0	4.2.0	
1	21.2.0	4.2.0	
1	21.2.2	2.1.17	Hedges
1	10.2.1	0.2.1	
1	3.0.0	0.3.0	
1	3.0.0	0.3.0	
1	3.0.3	0.3.2	

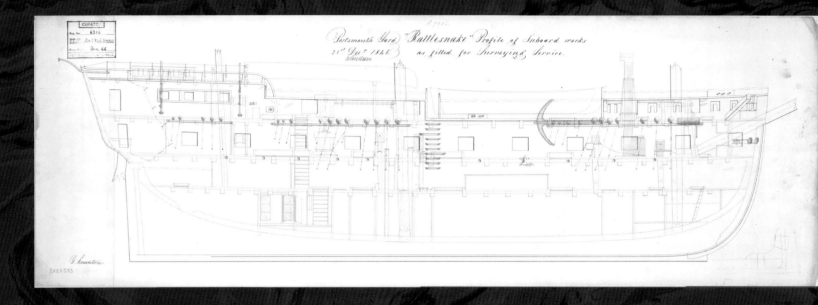

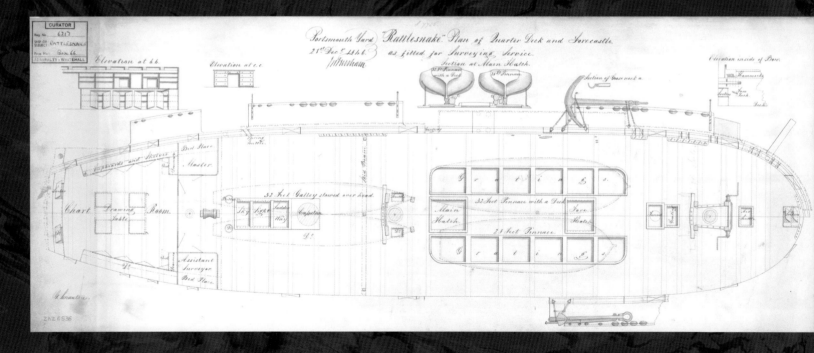

Inboard profile and plan of the quarterdeck and forecastle of HMS
Rattlesnake, as fitted for surveying at Portsmouth Dockyard in 1846.
(© National Maritime Museum, Greenwich, London)

intellectual precedence regarding his ideas on the origin of species. Hooker was appointed director of Kew in 1865 and remained Darwin's confidante for the rest of his life.

Also listed officially as an assistant surgeon, the brilliant and largely self-educated naturalist Thomas Huxley came to Australia aboard HMS *Rattlesnake* on a four-year expedition continuing the work of HMS *Fly* surveying the Great Barrier Reef, Torres Strait and southern New Guinea between 1846 and 1850. A much larger vessel than the *Beagle*, the *Rattlesnake* was home to 180 officers and men under the command of Captain Owen Stanley.

Like Hooker, Huxley was determined to advance his scientific career and while the official naturalist aboard the *Rattlesnake* was John MacGillivray, Huxley used the voyage to develop a scientific reputation as a highly talented anatomist and marine biologist. Encouraged by Owen Stanley, he was successful in having a paper published in the Royal Society's *Transactions* during the voyage.

Four years after his return from the *Rattlesnake* voyage, Huxley was elected a fellow of the Royal Society and appointed Professor of Natural History at the Royal School of Mines. During his career he rose to become, variously,

Hunterian Professor at the Royal College of Surgeons, president of the British Association for the Advancement of Science, and president of the Royal Society. However, it was as a pugnacious and very public advocate of Darwin's great work *On the Origin of Species* that Huxley famously earned the epithet 'Darwin's bulldog'. Like Joseph Hooker, Thomas Huxley remained a close friend of Charles Darwin throughout his life.

Charles Darwin was undoubtedly one of the great scientific figures of the nineteenth century, but his ideas did not develop in a vacuum. He lived at a time when the globe was opening up to science and his own experiences were supported by evidence gathered from many sources. Included amongst these were an Australian web, linking ships, surveyors and naturalists in a widening stream of information. Charles Darwin's genius lay in his ability to draw together the puzzling pieces of the jigsaw and to see, for the first time, the great natural panorama of life. Knowledge gleaned from southern lands and oceans was, and continues to be, a vital element of that vast panorama.

HMS *Bramble* at anchor in the Louisiade Archipelago.
(Oswald Brierly: Australian National Maritime Museum)

Arrival of first white men in the Louisiade Archipelago.
(Oswald Brierly, watercolour, 1860: National Library of Australia)

art and exploration

Richard Neville

Mitchell Librarian,
State Library of New South
Wales

The first Pacific voyage of Captain James Cook was one of the earliest to employ official expedition artists. This precedent, funded by collector and naturalist Joseph Banks, established a pattern of exploration art that flourished for the next 100 years. The value of the visual record in supporting textural descriptions was well understood long before Cook's voyages: what developed at the end of the eighteenth century was a wide-spread appreciation of the importance of professionally created and composed images, integrated with supporting text. The British Admiralty noted this when appointing John Webber as artist to Cook's third voyage, requiring him to 'make Drawings and Paintings as may be proper to give a more perfect Idea thereof than can be formed by written descriptions only'.[1]

The collections of the Mitchell and Dixson Libraries, and indeed the State Library

A toopapapoo of a chief. (John Webber, hand-coloured etching, 1789:
Mitchell Library, State Library of New South Wales)

of NSW, are plentiful in the work of expedition artists, travellers, professional colonial artists and amateurs. They are also rich in the published, and often illustrated, accounts of explorers and colonists, and a wide variety of plate books – a legacy of the regular stock of any large nineteenth-century public library. Indeed the growth in expedition and travel literature paralleled the rise in interest in, and was supported by, plate books of picturesque landscapes, significant buildings and antiquities of England, Europe and Asia. Together these images – original and published – comprise an abundant record of the colonisation of Australia and the Pacific region.

The importance of an expedition artist

The skills that trained artists could bring to their subject – their capacity to turn a landscape into a work of art, to make precise drawings of natural history specimens, and record recognisable profiles of important navigational features – were valued by travellers and explorers, natural philosophers and armchair critics. The illustrated book, or the collection of plates, of new or unusual countries and of their inhabitants, was a critical component of the intellectual process of European engagement and colonisation of the new worlds which began to be revealed in the nineteenth century. The quiet elegiac classicism and nascent exotic picturesque vision of the Pacific created by the artists William

Hodges and John Webber on Cook's second and third voyages were critical, for example, in shaping late eighteenth- and early nineteenth-century attitudes to the region. Their perception of the Pacific, promulgated in the accounts of Cook's voyages, was supported by publications of sets of 'views' such as John Clevely's four aquatints *Views in the South Seas* (c 1789) or Webber's own 16 etchings, also called *Views in the South Seas* (1786–1792).

Much later and minor publications, such as Conway Shipley's *Sketches in the Pacific: The South Sea Islands* (London, 1851), which included lithographs of the infamous Pitcairn Island, were hardly part of the great discourse of Pacific discovery, but they too are evidence of a persistent interest in, and market for, the exotic.

Viewing the art of science

It must be remembered, however, that images of science and exploration were just one part, albeit an important one, of a larger discourse, and that these images need to be set into the context of more wide-ranging discussions, observations, correspondence, text and print. Images are not simply static, faithful and reliable witnesses of the things they depict, but rather encapsulate a particular moment of looking. Explorers often travelled with the texts of earlier expeditions, and were well aware of, and keen to test, any existing iconography. George Stubbs' famous illustration of the kanga-

The chapel and schoolhouse on Pitcairn Island. (Conway Shipley, watercolour, 1848: Mitchell Library, State Library of New South Wales)

roo, 'An animal found on the coast of New Holland called kanguroo',[2] painted in England and modelled inaccurately on a stuffed animal, is often interpreted as a symbol of a European inability to 'see' Australian fauna. Yet when Comte de La Perouse observed a kangaroo in Botany Bay in 1788, he commented to Watkin Tench that Stubbs' engraving was 'correct enough to give the world in general a good idea of the animal, but not sufficiently accurate for the man of science'.[3] Similarly, Spanish botanist Antonio Joseph Cavanilles thought that the plates of Aboriginal figures, such as 'Natives of Botany Bay', in Arthur Phillip's *The Voyage of Governor Phillip to Botany Bay ...* (1789), failed completely to be true to reality, because the figures drew on Roman and Greek classical art rather than actual observation.[4]

While expedition art projects an ethos of direct reportage, and much of it was putatively about the recording of new encounters, the experience was rarely fixed and immutable. It could change, influenced by fashion, memory or particular agendas. Evidence of the flexible facts of documentary images can be seen in George Tobin's album of illustrations of the *Providence* expedition which successfully transplanted breadfruit from Tahiti to the West Indies in the early 1790s.[5] Tobin, an artist of some ability, was aware of the power of the image to convey information. Writing of the dolphin fish, he noted:

> I should say nothing of a fish so often described as the
> dolphin, but that the representations of it, even in the very best

prints to books of natural history, are erroneous. Perhaps the accompanying drawing ... will give you a better idea of its shape and colour than any thing that can come from my pen.[6]

Yet the care with which Tobin took to construct and compose his landscape images becomes clear when comparing the content of this album with a recently acquired collection of what appear to be preparatory watercolours.[7] On the subject of Tahitian funerary customs, for instance, comparing his album's 'A Toopaporo of Otahytey, with the corpse lying in it covered' with the earlier version 'A Toopapow, with the Corpse on it' shows how he enlivened the later watercolour by including a European gentleman and an unconvincing Tahitian woman. Curiously the mountain backdrop is replaced with a glimpse of sea, and a little stone sculpture, tucked away in the first version, is made more prominent in the second. On their own, either version could be considered an image of record – even if embellished – but in comparison it becomes clear that neither can be considered definitive.

Of course this is not to say that Tobin's watercolours are untrustworthy, but rather that, along with all images, the context, history and provenance of them need to be taken into account when assessing their evidentiary value. That many of the landscape watercolours in Tobin's album are on paper watermarked 1811 is clear evidence that they were compiled many years after the events they depict, and probably reflects a retrospective romanticisation of the expedition.

'A Toopaporo of Otahytey, with the corpse lying in it covered'. (George Tobin, watercolour, c 1792: Mitchell Library, State Library of New South Wales)

A Toopapow, with the Corpse on it. Island of Otahytey — 1792. Page 147.
WB.

'A Toopapow, with the Corpse on it'. (George Tobin, watercolour,
c 1811: Mitchell Library, State Library of New South Wales)

The historical value of the illustrated collections

The Mitchell Library recognised the value of Tobin's journal when it was purchased in 1915. However for many years before this New South Wales institutions had actively acquired images of the colony's history, encouraged in part by 1888 Centenary celebrations of settlement. In 1887, for instance, the Public Library of NSW (the former name of the State Library of NSW) purchased an album of 1790s watercolours of Australian birds.[8] Also in the late 1880s, the NSW Agent General in London, Sir Saul Samuel, negotiated the acquisition of a substantial tranche of Captain Cook material, which included a collection of William Hodges' drawings and an album of ornithological watercolours possibly by Georg Forster, as well as a number of relics and objects.[9] While the collection was first deposited with the Australian Museum, and not transferred to the Mitchell Library until 1955, its initial purchase paralleled the growing interest in Australiana of Australian collectors such as David Scott Mitchell and Alfred Lee. Mitchell's pursuit of Australiana began in earnest in the 1880s, and focused on printed and published material rather than manuscripts or pictures.

For many years after its opening in 1910, the Mitchell Library – built on the 1907 bequest of David Scott Mitchell – was considered as a de-facto national library. Other institutions, such as the Art Gallery of New South Wales, consciously transferred what it considered to be historical images to the library. Thus, for example, Charles Rodius' 1844 collection of fine Aboriginal portraits came to the Mitchell Library in 1921 from the Art Gallery on the grounds that their essentially documentary aesthetic was better suited to a historical collection. Similarly the Mitchell's important collection of Owen Stanley watercolours from HMS *Britomart* and HMS *Rattlesnake* were transferred from the Gallery in 1912 and 1921 respectively.[10]

In 1919, having realised that Mitchell's collection was still not strong pictorially, Sir William Dixson offered his historical pictures to the library. The first of many gifts was eventually accepted in 1929, and the residue of his collection was bequeathed to the library in 1952. The bequest included a number of single drawings by Hodges and Webber, and an album of 46 Webber watercolours.[11] Both David Scott Mitchell and Sir William Dixson were keen to acquire any published volumes of exploration – illustrated or not – and the library now holds a remarkably comprehensive collection. (It is also worth noting the often superior condition of the Dixson copies: it seems Sir William valued condition more so than Mitchell.)

Brierly on the *Rattlesnake*

Of the artists David Scott Mitchell collected, two who interested him above all others were also at one time shipboard artists: Conrad Martens, who sailed for a short time with FitzRoy and Darwin on the *Beagle*, and Sir Oswald Brierly.

A view of Huaheine. (John Webber, watercolour, c 1780: Dixson Library,
State Library of New South Wales)

Prince of Wales's Island canoe 'Bruwan'. (Oswald Brierly, pencil and wash,
c 1848 from *Sketches on Board the HMS* Rattlesnake: Mitchell Library, State
Library of New South Wales)

Brierly's early career was spent in New South Wales and the Pacific. He arrived in New South Wales in 1841 on Benjamin Boyd's yacht *The Wanderer* and was employed by the entrepreneur to manage his various enterprises at Twofold Bay. In 1848 he was asked by Owen Stanley to join the *Rattlesnake* expedition to the Torres Straits.

Brierly's technical skill is particularly evident in his depictions of Pacific sailing craft, a fascination he shared with the *Rattlesnake*'s naturalist Thomas Huxley and Stanley himself, but he was also a fine watercolourist capable of subtle and atmospheric interpretations of the Pacific landscape. His excitement in the expedition is palpable – he noted in his journal in June 1849 that he had an 'irresistible urge to sketch'.[12] Unfortunately this rich archive was not incorporated into any of the *Rattlesnake* publications, as Brierly did not return to England in time for their inclusion.

Stanley was also a compulsive sketcher, eager to record all the places he visited on both the *Britomart* and the *Rattlesnake* – from the pyramids of Egypt through to Singapore, Rio de Janeiro and Sydney. Yet his watercolours also document another side of expedition life, depicting a more 'domestic' view of shipboard life, with many drawings of social events and its daily activities. Images of material culture encountered by the expedition are recorded, but so too are his men setting up an observation post or enjoying a cooling swim while becalmed. Stanley was perhaps more alive to the mood in English art for genre painting, and the depiction of contemporary life in art.

In some respects Stanley's art parallels that of another one-time *Beagle* artist Augustus Earle, who was similarly fascinated by observing people, as can be seen in his 'Crossing the line'. It seems clear that Brierly, Stanley and Huxley collaborated on the voyage as the three often chose similar subject matter. Brierly's omnivorous documentation of natural history, material culture, coastal profiles and even languages, which he recorded in his journals, no doubt in part reflects the interests of his companions.

Naval artists in the South Seas

Drawing was also considered a useful talent for naval men, particularly for practical purposes such as delineating coastal profiles. Some officers, such as William Bligh, were merely competent. Others, like George Tobin, Owen Stanley (who had been tutored by family friend Edward Lear) or the resolutely self-taught Thomas Huxley, who served as assistant surgeon on the *Rattlesnake*, were blessed with natural ability.

While Huxley was skilled at dissection drawings of specimens he collected, he was also a capable landscape artist, and was no doubt influenced by Brierly. Lithographs made after his drawings were published in John MacGillivray's *The Voyage of HMS* Rattlesnake (London, 1852). The Mitchell Library's pencil drawing 'Party landed from HMS *Rattlesnake*' is derived from, but not identical to, 'Cutting through the scrub' which was published in McGillivray's *Voyage*.

CROSSING THE LINE.

Published by Henry, Colburn, Great Marlborough Street, 1838.

Crossing the line. (Augustus Earle, in R FitzRoy, *Narrative of the Surveying Voyages of His Majesty's Ships* Adventure *and* Beagle: State Library of New South Wales)

'Party landed from HMS *Rattlesnake*.' (Thomas Huxley, pencil: Dixson Galleries, State Library of New South Wales)

It is interesting to note that the pencil drawing was presented to the library by a King family descendant, and it is possible that Huxley gave it to Phillip Parker King in 1849. Similarly the library's watercolour 'Natives on the New Guinea Coast' – probably the village of Tassai – was presented to the library by a descendant of Colonel George Barney, of the Royal Engineers, who was also in Sydney in 1849. Presumably these two drawings were selected as gifts for Sydney friends.

Differing national approaches to art and science

Exploration illustration was in part framed by the nationality of the expedition and the agenda which shaped its activities. The numerous French voyages into the Pacific from the late eighteenth century onwards, which were born out of a blend of imperial ambition, a hope of commercial benefit and genuine scientific curiosity, often carried large scientific parties. Nicolas Baudin, for instance, travelled with what proved to be an unworkable 22 scientists and three artists. The resultant publications were a matter of national pride, and were financed by the French government. Similarly the United States Exploring Expedition, which circumnavigated the world between 1838 and 1842, was promoted as a national enterprise and a matter of pride for such a young country. While the expedition's focus was as much commercial as scientific, its altruistic contributions to science were intended to demonstrate to its supporters the cultural maturity of the

'Cutting through the Scrub.' (Colour lithograph from drawing by Thomas Huxley, in John MacGillivray, *Narrative of the Voyage of HMS* Rattlesnake, 1852: Mitchell Library, State Library of New South Wales)

United States. The expedition publication – five volumes of text and one atlas of illustrations provided by the official artists Joseph Drayton and Alfred Agate – was seen too as a statement of national accomplishment.[13]

On the other hand, the British Admiralty, particularly as the nineteenth century progressed, was less willing to subsidise artists and naturalists. The patronage and influence of Sir Joseph Banks was critical for Matthew Flinders being permitted to carry landscape and natural history artists. Yet arrangements for scientific positions and publication expenses on later expeditions were very much at the initiative of the commander. While the Admiralty was prepared to sanction and victual artists and naturalists, it made it clear that their interests were subservient to the real and commercial purpose of most voyages, which was to survey uncharted or poorly charted waters.

It was up to Robert FitzRoy, for example, to engage an artist for his second *Beagle* expedition. Knowing that none of his party could draw satisfactorily, he employed Augustus Earle as his shipboard artist. FitzRoy's Admiralty instructions were clear. His task was to survey the South American coast: natural history was a by-product of, rather than central to, the voyage. Indeed his instructions warned him against wasting time on 'elaborate drawings': 'Plain, distinct roughs, every where accompanied by explanatory notes' were more appropriate than 'highly finished plans, where accuracy is often sacrificed to beauty'.[14]

The publication of FitzRoy's expedition, *Narrative of the Surveying Voyages of His Majesty's Ships* Adventure *and*

Natives on the New Guinea Coast. (Thomas Huxley, watercolour: Mitchell Library, State Library of New South Wales)

Beagle *between the Years 1826 and 1836*, reflects this more modest approach. An octavo publication, illustrated with engravings interfiled into the text in the fashion of the day, its plates carry nothing of the gravitas of the grand expedition atlas. In part this is because the two expedition artists – Augustus Earle and later Conrad Martens – were more familiar with the genre of popular book illustration than the expanse of the large and formal landscape plate typically used in earlier expedition atlases, and still favoured by French expeditions. The illustrations contributed by Martens to the publication, such as his 'Patagonians', show an interest in atmosphere and the picturesque rather than the topographical literalness preferred by the Admiralty. Indeed his South American sketchbooks are often annotated with colour notes and descriptions of atmosphere.[15] Yet it is also true that his often sensitive, light-filled South American watercolours are ill-served in their translation into print.

Patagonians. (Conrad Martens, from R FitzRoy, *Narrative of the Surveying Voyages of His Majesty's Ships* Adventure *and* Beagle: Mitchell Library, State Library of New South Wales)

Sep.r 24.

a dark sky. warm gray. coming down flat and unbroken to the horizon lighter & broken in the upper part. the sun light up above the centre of the picture. — the sea not very rough, but sufficient to shew a squale, and worked out only by lights upon a ground of exactly the same tone and color as lower part of sky. — lights — the reflection of the sun streamed at the horizon with bright lights thinly scattered as they approach the foreground. and a very few touches of stronger dark to effect the appearance of the water — a schooner alone reefed —

necessary a part of the horizon may be relieved by washing back. the lower part of the sky, but this must be done very sparingly and at one side only. — the upper should be all light — make the gray with indigo, Ind.n Red. and yellow —

The *Beagle* at sea. (Conrad Martens, pencil and manuscript:
Dixson Library, State Library of New South Wales)

Depicting
indigenous peoples

Similarly influenced by the emerging French interest in the comparative study of human kind, French artists emphasised indigenous portraiture. Piron, Nicolas-Martin Petit, J Alphonse Pellion or Jacques Arago, for example, all made carefully observed, sometimes highly romanticised and often sympathetic portraits of Aboriginal people. By contrast, accounts of English expeditions, such as those of the *Beagle* and the *Rattlesnake*, while also concentrating on the indigenous peoples they encountered, present a much less formal visual record in their publications. In part this is

'Native, Twofold Bay.' (Oswald Brierly, watercolour, c 1843: Mitchell Library, State Library of New South Wales)

because of the often more modest size and ambition of English publications, but it also indicates the different agendas behind the various expeditions. The elaborate and expensive atlases of Jules Dumont d'Urville's *Voyage de la Corvette* l'Astrolabe, published in 1833, promote an entirely different view of exploration, one still infused with the grand rhetorical gestures of the heroic age of late eighteenth-century expeditions.

By contrast, the vigour of the sketch made from life can be startling. Brierly's quick and energetic pencil field sketch 'Native, Twofold Bay' presents very differently to the lithograph, after Louis de Sainson, 'Baie Jervis, Nouvelle Hollande' – which was published in *Voyage de la Corvette* l'Astrolabe: *Atlas Historique*. While it is easy to read Brierly's drawing as a 'truthful' representation of Indigenous people, de Sainson's more contrived image needs to be seen in the context of its overall publication: another, but different, artefact of colonisation.

Again, context should never be forgotten. One of the rarest works relating to the *Beagle* expedition is the separately issued lithograph by an unidentified, presumably English, artist titled 'The Three Fuegians brought to England by Captn Fitz-Roy, 1831'. (The second *Beagle* expedition was in part driven by the desire to return these three surviving Tierra del Fuegians, taken to England by the first *Beagle* voyage, to their homeland.) This image is about the direct power of civilisation: these neat individuals are celebrated for the impact of a Christian education – a consequence (even if ultimately unsuccessful) of the imperial agenda.

Pl 34.

de Sainson pinx.   Tastu Editeur   Lith. de Langlumé   V. Adam del.

BAIE JERVIS.
(Nouvelle Hollande.)

Les marins de l'Astrolabe partagent leur pêche avec les Naturels.

'Baie Jervis, Nouvelle Hollande.' (After Louis de Sainson, from *Voyage de la Corvette* l'Astrolabe: *Atlas Historique*: State Library of New South Wales)

THE THREE FUEGIANS.

Brought to England by CAPT: FITZ-ROY. 1831.

The Three Fuegians brought to England by Captn Fitz-Roy, 1831.
(R Day, lithograph: Mitchell Library, State Library of New South Wales)

Images of exploration are never neutral or unencumbered. Subtexts and unspoken agendas shape their vision and our interpretation of them. The pictorial and printed collections of the Mitchell and Dixson Libraries – which are much more encompassing than has been possible to describe here – offer unrivalled insights into the processes of colonisation and exploration from the earliest days of European movement into the Pacific region.

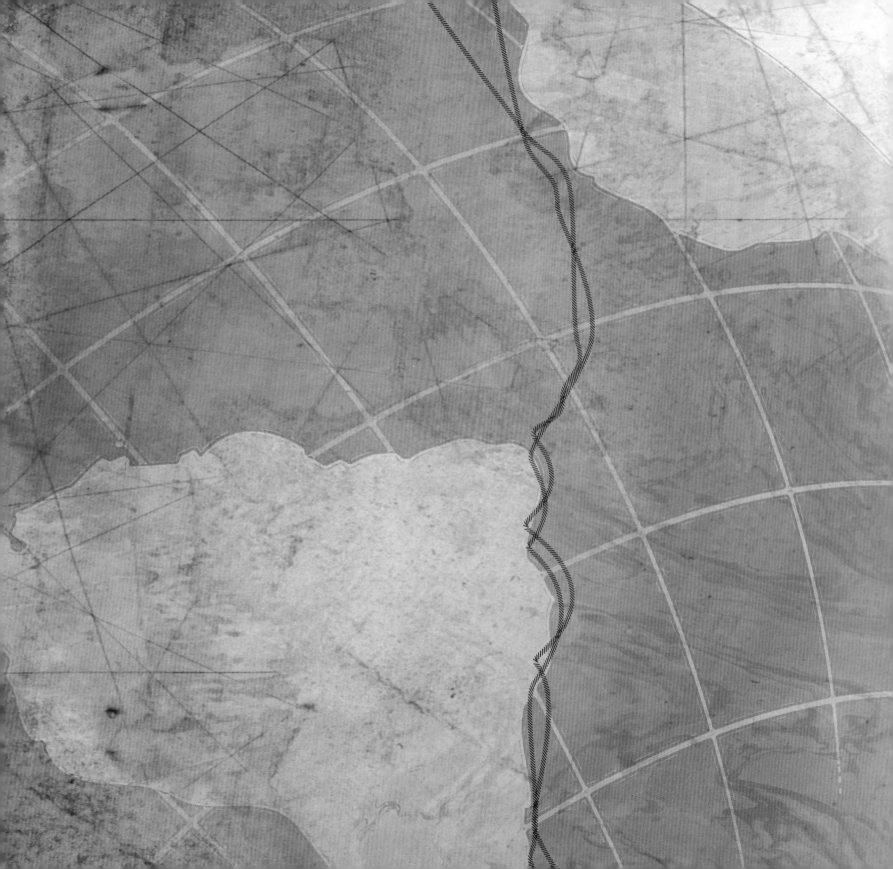

science on land and sea

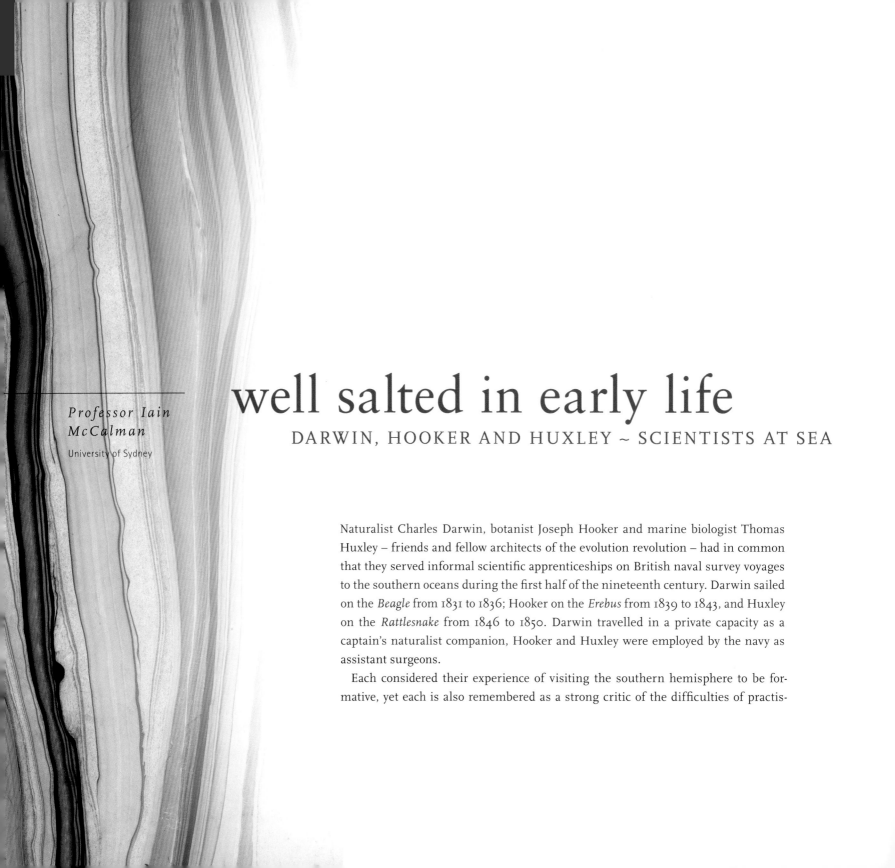

well salted in early life

DARWIN, HOOKER AND HUXLEY ~ SCIENTISTS AT SEA

Professor Iain McCalman

University of Sydney

Naturalist Charles Darwin, botanist Joseph Hooker and marine biologist Thomas Huxley – friends and fellow architects of the evolution revolution – had in common that they served informal scientific apprenticeships on British naval survey voyages to the southern oceans during the first half of the nineteenth century. Darwin sailed on the *Beagle* from 1831 to 1836; Hooker on the *Erebus* from 1839 to 1843, and Huxley on the *Rattlesnake* from 1846 to 1850. Darwin travelled in a private capacity as a captain's naturalist companion, Hooker and Huxley were employed by the navy as assistant surgeons.

Each considered their experience of visiting the southern hemisphere to be formative, yet each is also remembered as a strong critic of the difficulties of practising science while at sea. Of the three, Huxley advanced the most searching and systematic critique. In a review article of the *Narrative of the Voyage of the* Rattlesnake, published in the January 1854, he claimed that the culture and regimen of a naval survey vessel was incompatible with the practice of professional science.[1] Much of what he said was echoed in the more private letters and journals of Darwin and Hooker. Despite this, Darwinian evolution owes that wooden world a considerable debt. During the course of their voyages, and still more in later life, Darwin, Hooker and Huxley came to believe that the benefits of being at sea on British naval survey ships far outweighed the drawbacks.

The sorrows of young philosophers at sea

The initial and mutual incomprehension between a land-lubber scientist and the anti-intellectual seamen of a naval frigate has been well captured in bestselling novelist Patrick O'Brian's fictional figure of surgeon-naturalist Stephen Maturin, whose shipboard ignorance and dislike of naval discipline provokes the wry tolerance of the sailors. Darwin, Huxley and Hooker, none of whom had any experience of sailing ships prior to embarkation, faced similar initiations.

Darwin admitted having boarded the *Beagle* in 1830 thinking of a ship as 'a large cavity containing air, water and food mingled in hopeless confusion'. A month out of harbour he screamed in exasperation in his diary: 'Oh a ship is a true

Charles Darwin. (George Richmond, chalk and watercolour drawing, 1840: by kind permission of the Darwin Heirlooms Trust © English Heritage Photo Library)

Joseph Hooker. (TH Maguire, lithograph, 1851: National Library of Australia)

pandemonium!' He puzzled at the obsession with naval rank that made the conversations of his officer colleagues in the gun room mess 'so devoid of interest'. The relentless misery of seasickness made him curse both the sailors' vehicle and its element: 'I hate every wave of the ocean with a fervour', he told his cousin, and to his family confessed 'I loathe, I abhor the sea and all ships which are on it'.[2] Conversely, few historians seem to have realised the element of satire in the nickname of 'Philos' (philosopher) given him by the *Beagle*'s officers and crew. Later naturalist voyagers associated the title with the crew's mocking condescension. It implied that one was 'at sea' in a double sense: literally, an intellectual who had embarked on the ocean, but also a landlubber baffled by the ways of wooden ships and naval command.

Huxley and Hooker, lacking Darwin's private means and privileged status, were struck especially by the inadequacy of their scientific resources and the difficult conditions under which they were forced to practise their science. Hooker was aghast that the official naval kit issued for botanical work consisted of 25 reams of paper, two botanical vascula and two Wards' cases for bringing back plants alive. Preserving jars and fluid were regarded as entirely superfluous, and empty pickle jars filled with naval rum had to serve as substitutes. Luckily, Hooker's family came to the rescue, funding Joseph's uniform, private stores, essential instruments, and natural history reference books.

Huxley was even worse off. The navy gave him no scientific equipment, and his family, to whom he was already in debt, were too poor to help. Even his tow-net for catching jellyfish had to be improvised from ship's bunting, and then only after he had been forced to gather specimens with a rice sieve. The ship itself, an elderly 'donkey' or 'jackass' class frigate built six years before his birth, was crammed with 180 people and so poorly refitted that water sloshed around his cabin floor for the whole of the voyage.

Working at science brought a fresh crop of difficulties for both the two young surgeon's mates. On the *Erebus*, Hooker had to spread out his plant collections in the middle of the sick bay or on the gun-room mess floor where all his colleagues tramped. Swarming cockroaches inflicted other kinds of damage. The humidity made it almost impossible to dry specimens without damaging them.[3] The ship's pitching motion and his weak eyes made microscopy and accurate sketching difficult, while botanical reference books for studying plants at sea were hopelessly inadequate.

But at least Hooker was also allowed to work in the captain's cabin. Huxley had to attempt intricate dissections of jellyfish, stingers and sea squirts with a microscope lashed to the table in the *Rattlesnake*'s gun-room mess, which also housed 22 midshipmen – most aged around 14 or 15. A zoo would have been quieter and more spacious. And thanks to the Admiralty's meanness, the only available reference books had been bought out of his own pocket. One of them, a century-old natural history by the Comte de Buffon, caused his guffawing adolescent charges to nickname the objects of Huxley's investigations 'buffons': 'What a precious pack have I to deal with in these precious messmates of mine', he wrote acidly in his diary only a few days into the voyage'.[4]

Thomas Henry Huxley in 1857. (From L Huxley, *The Life and Letters of Thomas Henry Huxley*)

A deeper problem was the incommensurate cultures of work on board a survey vessel: 'The practical shiftiness required by the sailor in his constant struggle with the elements is ... far apart from the speculative acuteness and abstraction necessary to the man of science', Huxley wrote. The officers' unthinking indoctrination with naval routine and discipline produced constant petty clashes:

> if you want a boat for dredging, 10 chances to 1 they are always actually or potentially otherwise disposed of: if you leave your towing net trailing astern, in search of new creatures, in some promising patch of discoloured water, it is, in all probability, found to have a wonderful effect in stopping the ship's way, and it is hauled in as soon as your back is turned; or a careful dissection waiting to be drawn may find its way overboard as a 'mess'.

Even the presence of a captain sympathetic to science, as Owen Stanley claimed to be, was no guarantee of a general spirit of co-operation.[5]

A different order of challenge faced all three naturalists in the later stages of their expeditions, all of which lasted between four and five years. During his long southern voyages of the eighteenth century, James Cook diagnosed the exhaustion and home-sickness that gripped many of his officers and crew after several years at sea as 'nostalgia'. He thought of it as an actual disease of the nervous system which could only be averted by strenuous physical diversions like rum-drinking and dancing. This nostalgic condition struck Darwin soon after the *Beagle* left South America, when he knew that his extensive opportunities to work on land were over. Suffering more acutely than ever from the effects of the Pacific's long rolling swell, he was impatient to return to the isles of his heart on the other side of the globe.

Trivial as it might seem to others, he later recollected, the voyager could not help being depressed after a prolonged period of sailing by 'the want of room, of seclusion, of rest – the jading feeling of constant hurry – the privation of small luxuries, the comforts of civilization, domestic society'. The Pacific ocean loomed before him like 'a tedious waste, a desert of water', its blue expanses useful only in giving him 'the time and inclination to measure the future stages of our long voyage of half the world, and wish most earnestly for its termination'.[6]

During Huxley's second voyage to the Barrier Reef and northern Australia, between April 1848 and January 1849, he came close to a nervous breakdown. Some of the reasons were personal, yet he himself blamed the horrific conditions experienced on a wooden survey ship in the tropics. Despite sailing over the richest outdoor laboratory of jellyfish, corals, and pelagic sea creatures in the world, his scientific work ceased. The blank pages of his diary reveal someone in a state of mental paralysis. While other naturalists threw themselves into collecting birds, fish and shells, Huxley huddled in his humid cabin, catatonic with ennui. 'Fancy for five mortal months shifting from patch to patch of white sand in latitude from 17 to 10 south, living on salt pork and beef, and seeing no mortal face but our own', he moaned in

Becalmed near the Line – "Hands to Bathe"

'Becalmed near the Line, "Hands to bathe".' Even when becalmed there was
constant activity and noise. (Owen Stanley, watercolour, 1846–1849: Mitchell
Library, State Library of New South Wales)

a letter to his sister. The small idiosyncrasies of his mess-mates became 'sources of absolute pain, and almost uncontrollable irritation'. He began to conduct his formal duties like an automaton and to retreat into worlds of fiction and fantasy. While the *Rattlesnake* glided over a naturalist's paradise, Huxley read Dante's *Inferno*. Buried in the foetid underworld of his cabin, he extracted a perverse pleasure from its realistic details of suffering. Insofar as he observed nature at all, it was from his bunk: 'my sole amusement consists in watching the cockroaches, which are in a state of intense excitement and happiness'.[7]

Hooker's crisis came less from boredom or exhaustion than an inability to do botanical work at sea, because the *Erebus* expedition's Antarctic voyages in search of the south magnetic pole took them into a world almost devoid of plants. While at their Australian base of Hobart in Tasmania, Hooker had became so engrossed with collecting local botanical species that he half-considered feigning illness and remaining there when the expedition made its second foray south.[8] Long-delayed news of the death of his elder brother and immense pressure from his botanist father not to waste time working on marine science exacerbated his misery. In the end, however, it was only his sense of loyalty to Captain James Ross that overcame this temptation to abandon ship.

All three naturalists experienced periodic, and sometimes acute, difficulties in negotiating their relationship with their captains. Huxley equated the position of a survey captain to that of 'a demi-god; a Dalai lama, living in solitary state'.[9] Darwin's fiery clashes with FitzRoy are well known. Not only

was that captain haunted by a fear that a strain of hereditary madness would lead him to suicide, he also possessed a violent temper and a propensity for long sulky silences. James Ross on the *Erebus* was a less difficult personality, though he could be egotistical and imperious. Hooker's main complaint was having to devote so much time to working on Ross's private maritime collections and to sketching landscapes for the captain's intended journal.

As usual Huxley's case was the most extreme of all. On their voyage to New Guinea (May 1849 to January 1850) he accused his captain, Owen Stanley, of being both a bully and a coward. Huxley – and his fellow naturalists on the *Rattlesnake* – believed that Stanley's paranoid fear of the native

The land with no plants: *Erebus* and *Terror* confront the South Polar Barrier. (JE Davis, watercolour: Allport Library and Museum of Fine Arts, State Library of Tasmania)

'Dance, Brumi Island, New Guinea.' Huxley's contact with the New Guinea natives was limited to shipboard visits, due to Stanley's fear of landing. (Owen Stanley, watercolour, 1846–1849: Mitchell Library, State Library of New South Wales)

peoples led him to avoid contact with the land at all costs, thereby preventing them from exploring one of the last land masses still unknown to Western science. It was only after Stanley suffered a fit at the conclusion of the voyage, to collapse and die in Huxley's arms, that the firebrand surgeon's mate realised the inordinate toll that 20 years of surveying had exacted on the conscientious captain.[10]

The skills and disciplines of the wooden world

Huxley's change of mind about Stanley mirrored a gradual shift by all three naturalists towards a more positive evaluation of shipboard life. Darwin became conscious of the beneficial legacies as soon as he began to write his narrative of the voyage. Historian Janet Browne suggests that his daily journal was influenced by the example of FitzRoy's daily naval log. Darwin also began at sea to adopt a new regime of self-discipline when describing, assessing, cataloguing, preserving and recording his findings. He admitted that the daily shipboard rituals of navigating, mapping, logging, sketching, surveying, depth sounding, and magnetic and meteorological observation instilled a new 'habit of energetic industry and of concentrated attention to whatever I was engaged in'.

This 'training' or 'drilling my mind', as he called it, had begun even before the ship departed. His cabin-mate Stokes taught him how to make magnetic readings with a dipping needle, Captain King instructed him in meteorology, and

Captain FitzRoy, HMS *Beagle*. (PG King, ink and wash, 1838: Mitchell Library, State Library of New South Wales)

FitzRoy showed him how to determine longitude by comparing the position of the sun at midday with the London time recorded on numerous chronometers. 'I find to my great surprise', he now wrote to his family, 'that a ship is singularly comfortable for all sorts of work. Everything is so close at hand and being cramped makes one so methodical, that in the end I have been a great gainer.' In a way the *Beagle* was a floating college, with a holistic variety of knowledge available at an instant, including the significant artistic talents of Augustus Earle and, after he left the ship, of Conrad Martens. Darwin became habituated to the post-noon rhythms of shipboard work: while FitzRoy filled his daily logs and made surveying calculations, Darwin wrote up his journal, transcribed from his small portable notebooks, or tagged, numbered and described his specimens.[11]

Although Hooker was already a highly skilled botanist thanks to training from his father, life at sea extended his scientific expertise in other disciplines and methods. Captain Ross put him in sole charge of the ship's tow and dredge nets, and asked him to monitor their rich daily tranches of infusoria, molluscs, salpae and other open water sea-creatures. He also had to dissect and draw this zoological haul. After completing some 100 drawings, he boasted of having been given responsibility for 'a new field which none but an artist could prosecute at sea'. He even made significant marine discoveries of his own. He found, for instance, that the strange luminous patches in the ocean usually attributed to electricity or phosphorous were actually produced by tiny living animals, and that minute, complex animal life, or protozoa, provided staple food for fish and other marine creatures at the extreme depth of 400 fathoms.[12]

Ambitious and shrewd, Huxley purposely selected scientific research suited to shipboard travel in the southern hemisphere. He focused on analysing the comparative structures, or morphologies, of jellyfish and other perishable surface-swimming marine creatures that were 'long known but very little understood'. While the *Rattlesnake* sailed the southern oceans, he was able to net and study some of the rarest sea species on the globe. No common plan was known to unite this zoological farrago, their morphological relationships remained a mystery. French comparative anatomist George Cuvier had in 1812 placed them all in a category he called *Radiata*, but it was really, Huxley thought, 'a sort of zoological lumber-room' that cried out for re-organisation.[13]

On the voyage, Huxley taught himself to become a brilliant comparative anatomist, summarising his greatest scientific achievement as 'the reduction of two or three apparently widely separated and incongruous groups into modifications of the single type'. Using a microscope, he revealed an underlying set of structural and functional characteristics, undetected by previous naturalists, which linked several outwardly dissimilar classes of free-swimming hydra and jellyfish.

Another of Huxley's brilliant discoveries centred on whether salps (sea squirts) were individuals or colonies, since their life-cycles encompassed both solitary and chain-like forms. Here, his personal experience of shipboard life may have been a direct inspiration. Railing against the fact

View of the midshipmen's quarters on board a ship-of-war. (Augustus Earle,
watercolour, 1820: National Library of Australia)

that 'there is no means of separating oneself; no possibility of avoiding one another' on a wooden ship led him to ponder the nature of individuality. Should the solitary form of the salp be regarded as the biological individual, rather like the lone naturalist at sea; or should the whole chain of salps be seen as the individual, rather like the colony of sailors that made up a naval ship? His answer was that the marine biologist should see the individual creature as a process rather than a static state. The salp encompassed a succession of forms within a life-cycle that proceeded from one point of fertilisation to the next.[14] This startlingly original idea eventually helped to bring him a gold medal from the Royal Society.

Sailing on the southern oceans also brought Darwin and Hooker vital data and insights by taking them to islands which served as living laboratories for studying what we today call biogeography and ecology. Islands, thanks to their relative geological recency, their isolation and their manageable scale, were ideal places to investigate such issues as the migration and dispersal of new species, those species' adaptation to changing environments through the production of local varieties, their struggles with newer invasions, and their environmental interconnections within a confined locality.

After Darwin had spent two weeks exploring the Cocos Keeling coral atolls in the Indian Ocean, for example, he wrote in his journal on 12 April 1836: 'I am glad we have visited these Islands; such formations surely rank high amongst the wonderful objects of the world. It is not a wonder which at first strikes the eye of the body, but rather after reflection, the eye of reason.' For him the eye of reason had not only

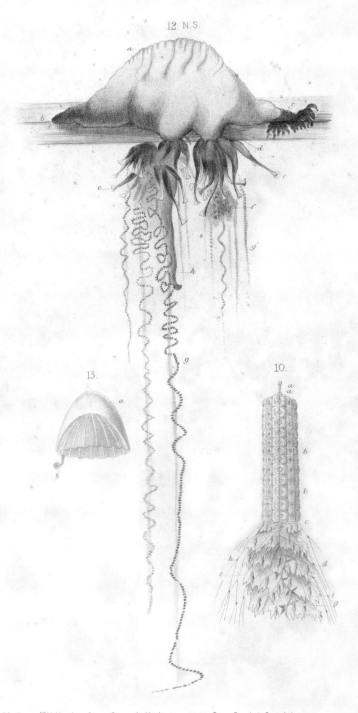

Medusa. (TH Huxley, from *Oceanic Hydrozoa*, 1859: Rare Book & Special Collections Library, University of Sydney)

confirmed his hypothesis that coral atolls were formed by the upward growth of coral polypi on a slowly subsiding seabed, but it had also revealed a new evidence about the functioning of what we would today call an ecosystem.

Darwin had seen how organisms influenced the environment and vice versa: tiny coral polypi actually created landforms on which new organic networks grew. Having collected what he believed to be a complete survey of Cocos biota, he discerned that the dispersal of a few common species on ocean currents from Australia and Sumatra, rather than a process of divine creation, had influenced the patterns of plants and creatures that found their way onto these remote isolated places. He noticed, too, clear evidence that organisms adapted their behaviour to suit new environmental conditions. Coral polypi had their growth limited and their shapes altered in accordance with the strength of the waves. He had even begun to notice and make use of metaphors of organic struggle to illustrate the dramatic and continual

Christmas Harbour, Kergeulen Island: Joseph Hooker's introduction to the mysteries of remote island floras. (JE Davis, watercolour: © National Maritime Museum, Greenwich, London)

Entrance of the Bay of Islands, New Zealand. While at the Bay of Islands Darwin remarked on the devastation wrought by the introduced Norway rat on the local flora and fauna. (Augustus Earle, watercolour, 1827: National Library of Australia)

battle between sea and coral: 'the ocean throwing its water over the broad reef appears an invincible enemy, yet we see it resisted and even conquered by means which would have been judged most weak and inefficient'.[15]

The freemasonry of the salt

The greatest debt of all three naturalists to life at sea was that it gave them a common culture and experience. Darwin was drawn to the young botanist Hooker in 1844 by their common 'general interest about southern lands'. He also needed access to Hooker's expert knowledge on the origins and relations of southern plant varieties and species, and on puzzles of botanical geography which might explain how some plants had managed to reach from the northern to the southern hemisphere, from the Antarctic islands to Fuegia, or from Australia to New Zealand, and yet develop distinctive new species in each place. His and Hooker's excited joint experiments immersing dried seeds in saltwater tanks to see if they would later germinate were part of the effort to solve such questions.

Southern sailing connected Charles Darwin equally to Thomas Huxley. When they first met in 1853, each was engaged in investigating the structural affinities and reproductive behaviours of marine species encountered on their respective voyages – Thomas Huxley with jellyfish, Charles Darwin with barnacles. Huxley's seaboard work on embryology found its way crucially into Darwin's later *Origin of Species*.

Ultimately the friendship between these three naturalists proved essential to ensuring the publication of the *Origin of Species* and to promoting it against the attacks of scientific, clerical and political opponents. Their powerful triangular relationship was underpinned by the fact that their sea voyages to the southern oceans had exposed them to common naturalistic challenges and interests. But it rested, too, on their shared experience as sailors who had braved exciting and dangerous situations, arduous physical and mental conditions, visited strange and beautiful places, and learned to respect and depend upon people far removed from themselves in social background and skill.

Darwin liked to think of himself and Hooker as 'co-circum-wanderers and fellow labourers'. There was no place for formality or affectation between them. Darwin's 'delightfully frank and cordial' sailor-like demeanour had struck Hooker the very first time they met. Hooker's personality had been similarly influenced by the salty camaraderie of the *Erebus*. He had made strong friendships with several of the junior officers, whom he later counselled on how to deal with grievances against Captain Ross and the Admiralty, and he never ceased to feel nostalgia for the days of 'having my own cabin at sea'. Darwin, for his part, saw in the young botanist all the character-shaping virtues that ship-board life could bring: 'good humoured patience, unselfishness, the habit of acting for himself, and of making the best of everything, or contentment: in short, ... the characteristic qualities of the greater number of sailors'.[16]

Cocos Keeling. (Photo Kim McKenzie)

A comparison of coral reefs around islands and atolls, including Keeling, or Cocos Atoll (fig 10) taken from FitzRoy's survey on the *Beagle*. (From Charles Darwin, *The Structure and Distribution of Coral Reefs*)

MAP SHOWING THE RESEMBLANCE IN FORM BETWEEN BARRIER CORAL-REEFS SURROUNDING MOUNTAINOUS ISLANDS, AND ATOLLS OR LAGOON ISLANDS.

'View of the Heads, Port Jackson. 1853' (Conrad Martens, watercolour, gouache, gum,
scraping out on paper, 54.2 × 76.2 cm: photo Diana Pannucio, purchased with assistance
from Overseas Containers Australia Ltd 1986, Art Gallery of New South Wales)

Darwin's sailor-like warmth and lack of social pretension had also won over the *enfant terrible* Thomas Huxley. Though itching to overthrow the corruptions of clerical and aristocratic influence that dominated 'the old guard' of British science, Huxley never saw the aristocratic Darwin in that light. Instead he felt that he shared with Hooker and Darwin a 'masonic bond in ... being well salted in early life'.[17]

Historian Janet Browne has observed that in the mid-1850s Charles Darwin began to conjure up the atmosphere and regimen of a 'ship on the downs' in order to re-enact that most productive period of his life on the *Beagle*. His study functioned like the *Beagle*'s cosy book-lined chart-room; wife Emma tended to his illness and protected his time like kindly Lieutenant Wickham had once fussed over the sea-sick

Philos; butler Parslow and the house staff functioned like his steward Syms Covington and the attentive crew; and the five surviving Darwin children played the part of the boisterous young 'middies', at once delightful and distracting.[18]

By setting his house in a chalky downland ocean 'on the extreme verge of the world', Darwin could live the protected, god-like life of a fleet captain. He could investigate barnacles under a microscope, just as he had once pored over the daily catch of the *Beagle*'s tow net; he could write up his notebooks with metronomic regularity in the manner of FitzRoy; and he could protect his privacy and peace of mind by funnelling relationships with the outside world through the medium of letters – apart from visits from his two fighting captains of evolution, Joseph Hooker and Thomas Huxley. The battle for the scientific theory of evolution was in a sense won as much at sea as it was on land.

Isolated Down House, where Darwin recreated shipboard life. (Photo Peter Fullagar)

Dr Jim Endersby
University of Sussex

a Gunn and two Hookers

FRIENDSHIPS THAT SHAPED SCIENCE

On Saturday, 10 October 1840, two men sat on the banks of the Derwent River, at Risdon, just outside Hobart, drinking bottled ale and discussing the gum tree they were sitting under. The older man was Ronald Campbell Gunn, Hobart's assistant police magistrate. The other was Joseph Dalton Hooker, assistant surgeon aboard HMS *Erebus*, a British naval vessel that had just arrived in Hobart after several months in Antarctic waters. The tree appeared to be a new species, so the two men – who were both enthusiastic botanists – collected several specimens from it. One of Gunn's specimens, its grey-green leaves and white flowers all dried to a uniform brown, is now at the Royal Botanic Gardens, Kew, just outside London, where it forms part of the herbarium – Kew's library of dried, pressed plants. Seven years after it was collected, Hooker gave the tree the scientific name, *Eucalyptus risdonii*, popularly known now as the Risdon peppermint gum.[1]

Given that Hooker's father, William Jackson Hooker, was director of Kew and that Joseph succeeded his father (becoming deputy director in 1855 and director in 1865), the presence of Gunn's specimen in Kew's collection is not surprising. But how did an Australian species come to be scientifically named by an Englishman who spent only a few months in the colony, rather than by the older and more experienced Gunn, who settled in Tasmania in 1830 and lived there for over 50 years? It was not as if Gunn was unqualified. As Hooker later acknowledged, Gunn collected 'with singular tact and judgement' and possessed 'remarkable powers of observation, and a facility for seizing important characters in the physiognomy of plants, such as few botanists possess'.[2]

The story of what happened to the twigs collected on that October day helps explain why Hooker rather than Gunn named the species, indeed why most of Australia's flora was named and described in London. The path that those gum tree specimens took also explains how Hooker eventually become one of the nineteenth century's most influential men of science: knighted for his services to empire, director of Kew and a close friend and champion of Charles Darwin. By contrast, Gunn is a minor footnote in the story of Tasmania's natural history.

It might seem as if this outcome was inevitable. Nineteenth-century Australia had no rivals to scientific institutions like Kew. Without collections and libraries, and access

The *Eucalyptus risdonii* specimen collected by Gunn in October 1840.
(© By kind permission of the Trustees of the Royal Botanic Gardens, Kew)

to scientific societies and their journals, men like Gunn could not hope to authoritively describe or classify their plants, whereas Hooker had all those resources and counted many of Britain's leading scientific men among his friends. But the story is still not quite that simple. When the gum tree specimens were collected in 1840, Kew had no library or herbarium – indeed the gardens were considered so shabby and ill-cared for that they were threatened with abolition.[3] Joseph Hooker had not yet built himself a scientific reputation, but was just a newly qualified junior doctor, and his father, William Hooker, had not yet been appointed as Kew's director. In 1840, it would have been difficult to predict the careers of Gunn, Hooker or the Risdon peppermint gum.

The 'great attraction'

Hooker's voyage aboard the *Erebus* began in September 1839, when the ship, accompanied by HMS *Terror*, set sail for Antarctica. Hooker was just 22 years old and it would be four years before he and the crew saw Britain again. Not long before he had set sail, Hooker had been given a present by a friend of the family who knew of Joseph's life-long passion for travellers' tales. It was a set of proofs of a still-unpublished book, Charles Darwin's *Journal of Researches* (now known as the *Voyage of the* Beagle). Hooker later recalled that while he was waiting for his own voyage to begin, he 'slept with the proofs under my pillow, that I might at once, on awaking, devour their contents'. He remembered that 'they impressed me profoundly, I may say despairingly', since Hooker felt that 'to follow in his footsteps, at however great a distance, seemed to be a hopeless aspiration'. But that was nevertheless what he hoped to do: to travel and publish his results, thus making a name for himself as a naturalist.[4]

Travelling was almost the only way in which people like Hooker and Darwin could establish themselves as serious men of science (women found it all-but-impossible). Apart from medical ones, there were no scientific degrees on offer in Britain until much later in the century, and there were almost no paid careers in science – not least because it was not really considered respectable to be paid to study nature.

For the previous two hundred years, science had been the exclusive concern of wealthy gentlemen, who would have scorned payment for their work. Gentlemen, by definition, did not work for money, so any form of compensation would have compromised their social standing. And the fact that they did not need the money was part of the reason they were considered reliable and honest investigators of nature – they were disinterested. By contrast, those who sought to profit from their discoveries, like the alchemists of earlier times, had every incentive to falsify their results, conceal their knowledge and exaggerate their achievements. The gentlemen who formed Britain's Royal Society in 1660 took pride in the fact that they received no funding from the state. Their wealth and social status guaranteed the accuracy of their observations and the truthfulness of their claims.

This ideal of the wealthy, disinterested scientific gentleman persisted well into the nineteenth century. In an ideal world,

Hooker would have liked to have been someone like Sir Joseph Banks, a wealthy landowner and friend of King George III, who accompanied Captain James Cook on his first voyage to the Pacific. Unlike Hooker, Banks had not received a salary from the navy. On the contrary, he had provided all the scientific equipment and personnel for the voyage, paying for everything out of his own admittedly deep pocket. On his return, his collections brought him fame and influence. He became de facto (but, of course, unpaid) scientific adviser to the government and effective director of the Royal Gardens at Kew.

However, Hooker (unlike Darwin) did not have a wealthy father who would leave him enough money to live on for the rest of his life. He needed to earn a living that would allow him to pursue his passion for botany. As he wrote in a letter to his father, 'I am not independent, and must not be too proud; if I cannot be a naturalist with a fortune, I must not be too vain to take honourable compensation for my trouble'.[5]

The *Erebus* and *Terror* expedition was not, of course, primarily concerned with botany. The ships were mapping variations in the Earth's magnetic field, which would allow compass readings to be corrected, thus making navigation safer. Although the ships had specially strengthened hulls, no wooden ship could withstand an Antarctic winter, so each year when the ice closed in they retreated north to places such as New Zealand and Van Diemen's Land (Tasmania), which allowed Hooker to collect plants in these still relatively unexplored regions. As he wrote to his father, 'No future Botanist will probably ever visit the countries whither I am going, and that is a great attraction'.[6]

It was while the ships were in Hobart that Hooker first met Gunn, who had been sending plants to Hooker's father for several years. The two men became friends immediately. Hooker wrote to his father: 'I have been nowhere save to my constant companion's Mr. Gunn, with whom I spend almost every night, & we have had several excursions together' adding that Gunn was 'a most excellent fellow, full of enthusiasm & cares for nothing but his plants'.[7]

Even though Hooker took advantage of his time ashore to travel and explore, collecting as many plants as he could, he did not have time to assemble extensive collections. Nevertheless, once he returned to Britain he decided that instead of simply describing his own collections he would produce something much more prestigious – the first comprehensive floras of the southern oceans. His books would be immense catalogues of plants, standard reference works for all future naturalists, but he had even more ambitious plans for them. By analysing the similarities and difference between the communities of plants of the various countries around the Antarctic circle, Hooker hoped to be able to discern patterns that would allow him to deduce the scientific laws that explained why plants grew where they did. Discovering such laws would enhance his own reputation, of course, but also that of botany itself. In the early decades of the nineteenth century, botany was looked down on by the practitioners of more prestigious and demanding sciences, like physics and astronomy. A key reason for its lowly status was that botany seemed to be little more than picking and naming flowers: it lacked the kinds of mathematically precise laws that characterised the 'higher' sciences.

However, understanding the laws of plant distribution was of more than purely academic interest. Britain's immense wealth was based largely on trade, and that trade was dominated by plant crops like cotton, tea, rubber, spices, timber and opium. Plants were one of the economic foundations of the British empire, so understanding how and why they grew in particular places was the key both to finding new crops and to transplanting economically valuable plants to new British colonies where they could be exploited profitably. Just as geologists had raised the standing of their science through a mixture of philosophical speculation and successfully predicting gold deposits, botanists like Hooker hoped to achieve similar advancement with a similar strategy. Supplying some laws would help both Hooker and botany gain a little more prestige in the eyes of Britain's leading men of science.

'Enthusiasts like ourselves'

To realise his ambitious plans, Hooker needed more plants. He could rely on his father's collections, but he would need many more dried plants to build up the truly global herbarium that his books would be based on. Naturally, he turned to friends like Gunn, who was just one of several dozen collectors to supply him with specimens. When he first formulated his plans, Gunn wrote to say that:

> Your father has delighted me with a proposal which he intends making to you to publish all the Southern plants! It will be a

magnificent concern. For my sake bash away at it. I shall send you my specimens so far as possible.[8]

During these early years of his career, Hooker had very little money. As he told Gunn, the Admiralty continued to provide his assistant surgeon's pay of £130 per annum, 'on which I grow uncommon fat as you may suppose'.[9] Given his lack of money, Hooker could not possibly pay men like Gunn for their work, but in Gunn's case, any offer of payment would have been refused. Although Gunn came from a very modest background (his father was a private soldier), it was comparatively easy to raise your social status in the fluid world of Britain's nineteenth-century colonies. Gunn had been introduced to botany by the son of a local landowner, who taught him to view the science as a gentlemanly pastime, a respectable hobby that was improving to the mind and contributed to the scientific progress of both colony and mother country. Gunn shared some of the same ideals as Hooker, but saw botanising as a way of becoming a gentleman – as long as it was pursued in a gentlemanly way, by not accepting payment for specimens.

Thanks therefore to Gunn's desire to be a gentleman, Joseph and William Hooker did not have to worry about their inability to pay for specimens. However, from the Englishmen's perspective, Gunn's social aspirations also had some drawbacks. When Hooker complained about the difficulty of earning a living from his science, Gunn responded that:

> Your account of the Rewards bestowed upon Science & learning in England is not encouraging – and it hardly

required your letters to satisfy me that Natural History must be followed for its own sake alone by enthusiasts like ourselves.[10]

Gunn's inclusive phrase 'enthusiasts like ourselves' shows that he saw himself as the Hookers' equal, not as a mere collector and certainly not as an employee. The plants he sent were valuable gifts, and he expected to receive similar gifts in exchange, especially of scientific instruments, botanical books and journals, which were virtually unobtainable in the colony.

Among the offerings Hooker sent after he returned to Britain was a short article on the 'Cider Tree' of Van Diemen's Land in which he described how the colonists produce 'a liquor resembling black beer, obtained by boring the trunk' of a previously nameless species of gum tree. The tree had first been mentioned in print by the Quaker traveller James Backhouse, but Hooker gave the details of how the 'cider' was extracted by quoting Lieutenant William Breton, whose journal describing his travels in Tasmania had been published in the colony's first scientific journal.[11]

In addition to describing the cider-making, Hooker gave the new species its scientific name, *Eucalyptus Gunnii*, to honour his Australian friend, who had known both Backhouse and Breton. Hooker sent Gunn a copy of the article, commenting: 'I have called the cider tree Gunnii ... I hope you will like my yarn about it – very romantic ain't it?'[12]

Although Gunn was indeed delighted by Hooker's compliment, he felt compelled to add that he:

Ronald Campbell Gunn, looking every inch the gentleman he aspired to be.
(© By kind permission of the Trustees of the Royal Botanic Gardens, Kew)

was much amused at your *quotation* from *Breton's* Journal – which was in fact furnished by me – from Breton *never having seen the tree*! He ought in justice to have quoted me as his authority for the tree existing – as also for the manner of collecting the Sap, but he lives only by borrowed matter. Half the *sense* of his Journal was furnished by me over a bottle of wine at Penquite – Breton having gone over the ground without even noticing what was the *Rock* of which the mountains are composed although he is a Fellow of the Geological Society. He is a good hearted fellow but a great fool.[13]

Gunn was clearly proud of his own expertise and understandably touchy about those who 'borrowed' it without acknowledgement.

Gunn's dismissive attitude to those like Breton who built their scientific reputations on 'borrowed matter' is understandable, but Hooker himself might have been equally accused of such reputation building, since he relied on the local knowledge of collectors like Gunn, Breton, Backhouse and many others. Without their assistance he could not have written the books that established his reputation. Given that Gunn did not want money for his specimens, Hooker had to find other ways of thanking him. Naming a species after him was one way (although, as the case of the cider tree illustrates, even that could backfire slightly), and sending botanical publications out to him was another.

Botanical books and magazines were, in many ways, the ideal way to thank a collector, since they gave him (almost all the collectors were also men) the pleasure of seeing his name in print. Such gifts also helped educate collectors: they not only improved their knowledge of botany and its terminology – which helped make their notes more accurate and concise – but as they learned which species had already been described and named in European publications, colonial collectors were also being instructed to focus on novelties. But as the colonial naturalists educated themselves, their confidence grew.

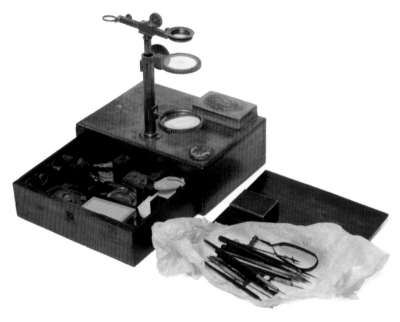

Joseph Hooker's dissecting microscope, together with some of the instruments used for dissection. Similar tools were all-but-unobtainable in early nineteenth-century Australia. (© By kind permission of the Trustees of the Royal Botanic Gardens, Kew)

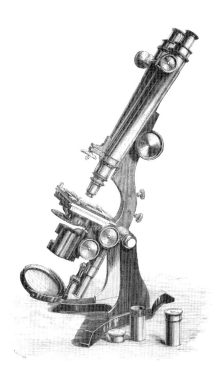

A perfect gift for a botanist: a large and expensive compound microscope. (The Science Museum, London)

Eucalyptus gunnii. (JD Hooker, *Flora Tasmaniae*, plate 26: © By kind permission of the Trustees of the Royal Botanic Gardens, Kew)

Plate XXVII

Fitch del et lith

Vincent Brooks Imp

Eucalyptus Gunnii , *H.f.*

'Only dried specimens'

When in 1836 William Hooker sent Gunn the latest issue of
his magazine *Icones Plantarum*, Gunn was 'much pleased'
with the pictures of his plants, but added that:

> they are very correct except Correa Backhousiana – the flowers
> of which are pendulous & not erect. They have been rubbed
> upwards in drying by the Gentn. who sent them to me from
> Woolnorth.[14]

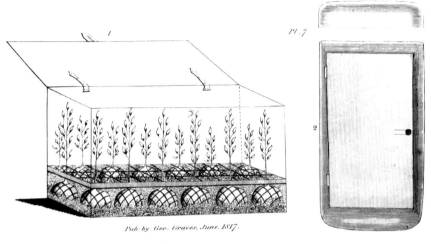

Pub. by Geo. Graves, June 1817.

A case for transporting living plants (left) and a 'vasculum' or collecting box.
(George Graves, *The Naturalist's Pocket-Book*, 1818, plate 7: Whipple Library,
Cambridge University)

By this time, Gunn had become a recognised expert in the
colony, and as a result other would-be botanists, like the 'gentleman from Woolnorth', sent him specimens. However, this
particular collector was evidently not as skilful as Gunn and
as a result, the specimen from which the plate was drawn had
been distorted. The artist who then engraved it (who would
never have seen a living specimen of the plant) simply copied
the poor specimen.[15]

Such mistakes were perhaps inevitable among those who
had not seen the living plant. By contrast, as Hooker acknowledged, 'there are few Tasmanian plants that Mr. Gunn has not
seen alive'.[16] Of course, what he did not say was that Gunn had
also seen more living Tasmanian plants than Hooker himself.
This was potentially a problem, since some colonial naturalists
argued that the dried, herbarium specimens Hooker used were
no substitute for first-hand knowledge of their living counter-
parts, not least because some features of the living plant were
simply not detectable in the dried specimen.

A plant press, 1852. (Sarah Crease, 1852: British Columbia Archives)

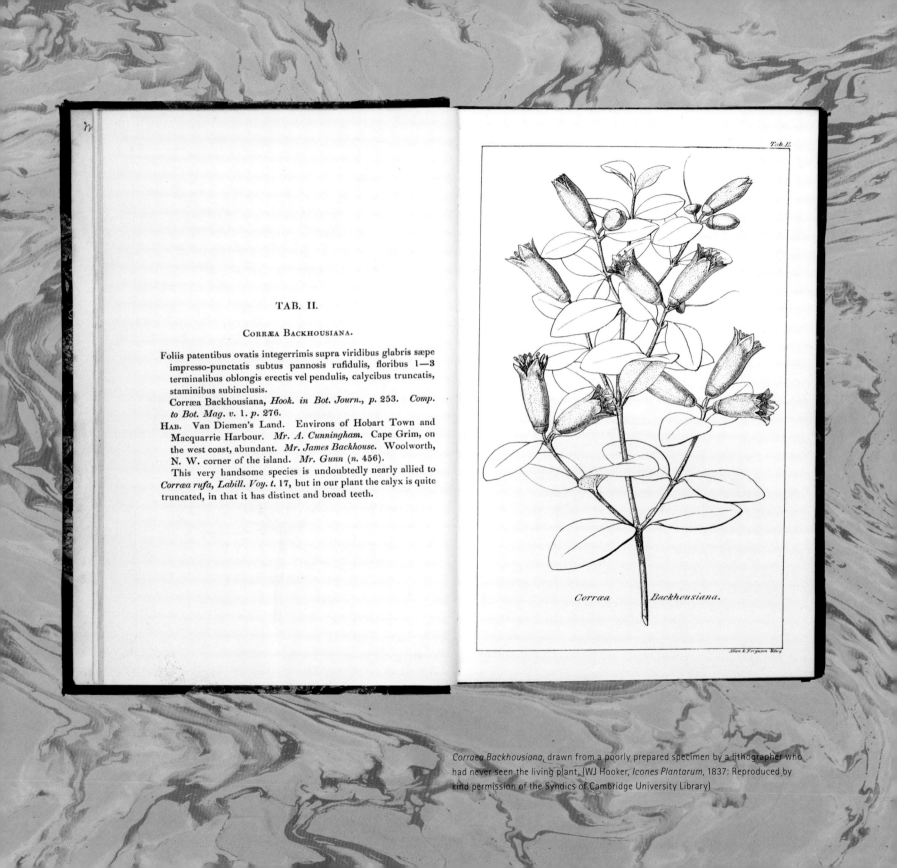

TAB. II.

Corræa Backhousiana.

Foliis patentibus ovatis integerrimis supra viridibus glabris sæpe impresso-punctatis subtus pannosis rufidulis, floribus 1—3 terminalibus oblongis erectis vel pendulis, calycibus truncatis, staminibus subinclusis.

Corræa Backhousiana, *Hook. in Bot. Journ.*, p. 253. *Comp. to Bot. Mag. v. 1. p. 276.*

Hab. Van Diemen's Land. Environs of Hobart Town and Macquarrie Harbour. *Mr. A. Cunningham.* Cape Grim, on the west coast, abundant. *Mr. James Backhouse.* Woolworth, N. W. corner of the island. *Mr. Gunn (n. 456).*

This very handsome species is undoubtedly nearly allied to *Corræa rufa, Labill. Voy. t.* 17, but in our plant the calyx is quite truncated, in that it has distinct and broad teeth.

Tab. II.

Corræa Backhousiana.

Allan & Ferguson, Libn.

Although he generally deferred to Hooker's expertise, Gunn's confidence in his first-hand knowledge led him to argue with Joseph Hooker about how many Tasmanian species of the genus *Tetratheca* there were. He commented, 'the Tetrathecas bother me a *little* I must confess, but I do not despair [of] proving by & bye that we have *at least* 4 species'.[17] His confidence was based on the fact that the genus – a group of small flowering shrubs, some of which look a little like heathers – is only found in Australia. Another Australian-based naturalist, Ferdinand von Mueller, told William Hooker that:

A good chara[cter] for distinguishing Tetrathecae is offered also by the direction of the sepals in a fresh state; I adopted it in my own diagnosis of Tetrath. baueraefolia, but neither Steetz nor Scuchhardt could make use of it as they saw only dried specimens.[18]

As both Mueller and Gunn realised, although metropolitan experts had greater access to books, the colonials had better access to the living plants: no European botanist could have examined more living plants than his colonial correspondents had.

In his letters, Gunn referred regularly to the importance of being 'on the spot' and seeing the plants alive. As he told Hooker:

Your rambles in Tasmania, limited as they have been will be of immense service to you in your future researches. In my future Communications you will be able to appreciate the remarks I may make about our Plants, their Soil, & habits. A Fern tree Scrub to wit could not be explained to one who had not seen it.

Helichrysum leucopsideum, collected by Gunn in 1841. (Collection: Tasmanian Museum and Art Gallery)

TASMANIAN HERBARIUM
HOBART   HO

12341

Tasmanian Herbarium: Hobart   HO 12341
Flora of Tasmania   **District:** North East
Family: Asteraceae
Helichrysum leucopsideum DC.

Collector: R.C. Gunn   **No:** 426
Lat: 41° 06' S   **Long:** 146° 50' E   **Date:** 03 Dec 1841
Map: **Grid:**   **Alt:**   **Prec:** 2
Locality: George Town

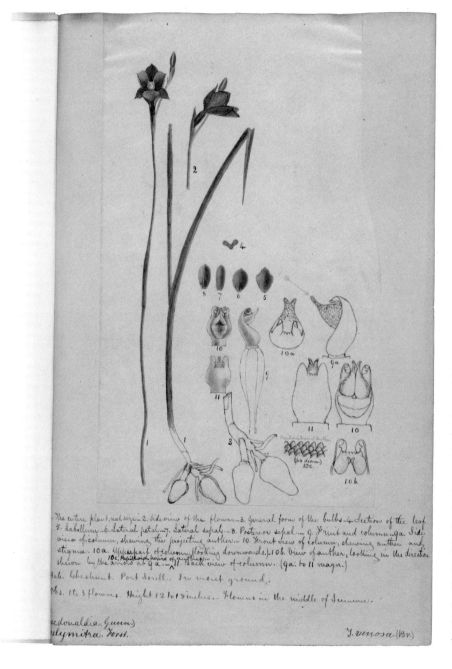

An orchid, painted from the living plant by Gunn's friend William Archer (By permission of the Linnean Society, London)

Gunn also reminisced in this letter about their collecting trips together, and asked if Hooker recalled 'drawing Corks out of *Bottled ale*'.[19]

Hooker remembered the trips, the beer and Gunn with great fondness. He replied, 'what would I give to have you here, old friend, for one little day? I often wonder how your game leg is; & whether bottled ale is good yet.' He also mentioned that he and his father were 'very desirous of a portrait of your august self if such a thing is to be got done out there'.[20]

Hooker was grateful for Gunn's friendly willingness to collect tirelessly (and without pay). He could not afford to alienate those on whose collections he depended, but nor could he accept their argument that descriptions of Australia's plants could only be based on living examples. Even Kew could not have hoped to grow more than a fraction of the over 8000 species Hooker described in his 'Essay on the Flora of Australia' (1860), so accepting their views would have meant surrendering the chance to write the prestigious and influential book he needed to advance his career.[21]

Instead, Hooker argued that it was *only* dried specimens that could be used to write a truly comprehensive flora, since they alone allowed a truly global collection to be assembled, thus allowing an accurate assessment of the relationships between the plants of different parts of the globe. In many cases, he simply overruled the colonials' opinions as to how many plants their countries possessed. Fortunately, being a gentleman also meant being polite, so Gunn did not generally dispute Hooker's judgments, but not every collector was as compliant, and Hooker had to engage in some fairly delicate negotiations with some of the other Australasian naturalists. In the case of the New Zealand naturalist, William Colenso, for example, Hooker helped him become a Fellow of both London's Royal and Linnean Societies, partly to defuse Colenso's irritation at Hooker's persistent refusal to let him name New Zealand's plants.

The 'throne of science'

Europe's museums and botanic gardens are full of dried plants, stuffed animals, beetles and bird skins, cabinets of fossils and boxes of bones. Men like Gunn, Colenso, Backhouse, Mueller and even the 'great fool' Breton, spent time, money and enormous skill to bring these treasures from all over the world, sometimes risking their lives in the process.

As specimens poured into the great cities of London, they began to assert an increasingly greater pull on distant parts of the world, sucking in more and more specimens as the fame of their collections and curators grew. Gunn had helped William Hooker build his herbarium and reputation, which in turn helped him secure the directorship of Kew when the government took over its management from the royal family in 1841. Further help from Gunn, together with his father William's collections and influential friends, helped Joseph Hooker build a career that brought him unprecedented influence.

In 1873, London's *Gardener's Chronicle* reported that Hooker had been elected president of the Royal Society. It was, their editorial commented, 'an event peculiarly gratifying to bota-

Hooker in 1877. He sent the photograph to Gunn with a note 'from his very old friend, JD Hooker'. (Ronald Campbell Gunn collection, Alexander Turnbull Library, Wellington)

From the ROYAL GARDENS, KEW.

nists and horticulturalists. It is many years since the throne of science has been occupied by a naturalist – we believe not since the time of Sir Joseph Banks.'[22] Hooker had finally succeeded in following in Banks' footsteps. But he could not have done so without the help of men like Gunn, as he acknowledged in his *Flora Tasmaniae* (1860), which was dedicated to Gunn and his fellow Tasmanian, William Archer. In his introduction, Hooker commented that 'I had the pleasure of making Mr. Gunn's acquaintance ... in 1840, and am indebted to him for nearly all I know of the vegetation of the districts I then visited'. He added that, 'I can recall no happier weeks of my various wanderings over the globe, than those spent with Mr. Gunn'.[23]

Gunn too had reason to be grateful to the Hookers, since his correspondence with them had brought him both treasured friendships and many gifts of books and journals. It was

 A Fern Valley, VDL.

'Fern Tree Valley, VDL', from a book by one of Gunn's friends that captures the colonial naturalist's isolation. (From James Backhouse, *A Narrative of a Visit to the Australian Colonies*, 1843: © By kind permission of the Trustees of the Royal Botanic Gardens, Kew)

Gunn's expertise in Tasmania's natural history that brought him to the attention of Sir John Franklin, who became lieutenant governor of Van Diemen's Land in 1837. Two years later, Gunn became Franklin's private secretary and then clerk of the executive and legislative councils. Thanks to the governor's patronage, Gunn also became secretary to the Tasmanian Society, which Franklin had founded, thus ensuring that he met all the scientific visitors to the island.[24]

Science created opportunities for Gunn, but in the late 1850s he seems to have gradually given up botany. He did some geological work for the colony's government, looking in vain for gold deposits to rival those of Victoria, and then settled near Launceston where he became a landowner and worked as recorder of land titles until ill-health forced him to retire in 1876. He died five years later, having finally become the gentleman he always aspired to be.

The work of colonial naturalists brought specimens, and ultimately the power to name and describe them, to London. However, not all Gunn's specimens of *Eucalyptus risdonii* remained in the empire's capital. On the corner of the one collected at Risdon on that Saturday in 1840 there is a note that reads 'A sheet of this number (1278) given to Mr. Maiden, Sydney Bot. Gardens Oct. 1900'. Gunn's few twigs had circled the earth and the specimen is still there, more than 160 years after it was collected, not merely a valuable tool for botanical researchers, but a tribute to a friendship that helped shape nineteenth-century science.

John MacGillivray

MERITS ALL HIS OWN

Sophie Jensen

Australian National
University & National
Museum of Australia

In order to commemorate the visit in 1854 of HMS *Herald* to Raoul Island, the north-ernmost island in the southern Pacific's Kermadec group, Captain Henry Mangles Denham chose to name various landmarks after some of his officers. The newly bestowed names included Rayner Point, Milne Islets, Parsons Rock and MacGil-livray Bluff. Seven years later, when Denham's charts arrived at the Admiralty in London, one name had been erased: MacGillivray Bluff had become Fleetwood Bluff, renamed after Denham's son.[1] In many ways this disappearance captures far more about the life of naturalist and collector John MacGillivray, for whom it was initially named, than had it remained as a commemoration of his visit.

Although in this instance MacGillivray's name may have been erased geographi-

THE EXPEDITION TO THE SOUTH SEAS.—H.M.S. "HERALD," AND THE "TORCH" STEAM TENDER.

The expedition to the South Seas, HMS *Herald* and the steam tender *Torch*.
(*Illustrated London News*, 15 May 1852: National Library of Australia)

cally, historically it is woven throughout the narratives and accounts of nineteenth-century collecting and exploration in the Pacific and Australia. His presence as assistant naturalist on board HMS *Fly* (1842–1846) and naturalist aboard HMS *Rattlesnake* (1846–1850) and HMS *Herald* (1852–1855), combined with his later career as a private collector, mean that few could rival his levels of knowledge and experience as a collector and ethnographer of Australian and Pacific material. The career path he followed, the choices he made, the opportunities he pursued, his difficulties and, sometimes spectacular, failures (and failings) all help to create a vivid picture of the intricacies of the system of natural history collecting during this crucial period.

A career dwarfed by giants

On the surface it would seem that MacGillivray had all the ingredients required to establish a successful and prominent career within the British scientific community. By the time he was appointed as naturalist on the *Rattlesnake* in 1846, it appeared that he was on track to do just that. Only four years older than Thomas Huxley, the assistant surgeon on board the same ship, he already had one major voyage under his belt and, unlike Huxley, a strong family background in the natural sciences.

The ghostly presence of MacGillivray Bluff, however, reflects a life eventually overwritten by the reputation and achievements of his contemporaries. As well as Huxley, these were

THE AUSTRALIAN ZOOLOGIST, Vol. ix.   PLATE IV

The only known picture of John MacGillivray, which accompanied his brother-in-law's article on his life 'A martyr to science'. (*Good Words*, 1868: National Library of Australia)

The *Rattlesnake* and *Bramble* finding an entrance through the reefs into the Louisiade Archipelago, 14 June 1849. (Oswald Brierly, lithograph from watercolour: National Library of Australia)

C.W.BRIERLY DEL. T.G.DUTTON,LITH.   London, Published July 8th, 1852, by Ackermann & Co. 96, Strand.   DAY & SON, LITHrs TO THE QUEEN.

Nº 1.

H. M. S. "RATTLESNAKE",
& BRAMBLE TENDER.

Commanded by Captain Owen Stanley R.N. finding an entrance through the Reefs into the Louisiade Archipelago, S.E. extreme, New Guinea, June 14th 1849.

To Rear Admiral Sir Francis Beaufort, K.C.B. D.C.L. F.R.S. &c. &c. This print is with permission respectfully dedicated by

The beacon built at Raines Islet during the voyage of HMS *Fly*. (Edwin
Augustus Porcher, watercolour, 1844: National Library of Australia)

some of the leading intellectuals, scientists, explorers and collectors of his day – figures such as Joseph Beete Jukes, Joseph Hooker, John Gould and Owen Stanley. The experiences and careers of many of MacGillivray's associates were crucial in the debates and development of Darwinian ideas in Britain and abroad. MacGillivray, however, is often regarded as never having lived up to the potential he first demonstrated as a young, enthusiastic assistant naturalist collecting for the celebrated Earl of Derby aboard the *Fly*.[2] He did not publish widely and made no significant contribution to the intellectual debates and discussions surrounding, and in many cases stimulated by, the collections he amassed. MacGillivray remained primarily a collector, observer and documenter.

In 1867, twenty-one years after he and Huxley had set out together on the *Rattlesnake*, MacGillivray died poor and alone in a cheap Sydney hostel, so removed from the society he had previously inhabited that his death certificate recorded 'mother and father unknown'.[3] At the same time, back in England Huxley was busy making his name as a key player within a dynamic scientific community and on his way to becoming one of England's most powerful intellectual figures. The stark contrast of these two lives, which ran for a short time along parallel lines, makes MacGillivray's life, career and eventual fall from grace, particularly intriguing.

Unlike many other visitors to the Antipodes, MacGillivray never settled back in Britain. This experience of dislocation and relocation is an important contributing factor to our understanding of his responses to the people, landscapes and environments throughout Australia and the Pacific. His attitudes evolved along with his status as he was first a visitor, later an expert and finally a resident. He is also representative of each of the major streams of scientific collecting during the period. He worked as a private collector for celebrated individuals such as the Earl of Derby, Hugh Cuming and John Gould. He was a key participant in the tradition of naturalist voyagers attached to British surveying vessels, and is also representative of that class of collectors who struck out on their own, as agents roaming in search of profitable ventures, saleable items and personal fortunes. MacGillivray's life and collections therefore provide a unique insight into the mindset and experiences of a collector and naturalist in the nineteenth century.

An unsurpassed fieldworker

There is one area, common and key to each of the phases in MacGillivray's career, in which his reputation has remained intact – his energy, zeal and skill as a fieldworker remain undoubted. It is difficult now to imagine the way in which a collector such as MacGillivray would have viewed the world. To see through such eyes would be to be constantly alert and searching. In a telling passage from a letter to a family friend, MacGillivray complained that in train travel 'the dazzling of the eyes by our swiftly dashing past objects which I continually strain my eyes to see produced a most painful sensation from which I have just recovered'.[4] Unlike a casual gaze of admiration or even the careful observation of an interested

The settlement at Port Essington. (Edwin Augustus Porcher, watercolour, 1845: National Library of Australia)

onlooker, MacGillivray's eyes were trained to seek, to find, and to analyse his environment in every detail and at every opportunity. His interests in zoology, ornithology, botany, ethnography and geology meant that his every sense would have been engaged, alive and every feature of the environment under constant scrutiny.

Each new landscape, island or environment was a naturalist's potential goldmine, every individual encountered a source to be exploited. Each birdcall overhead, insect fluttering by and track underfoot offered possible discoveries. MacGillivray would have looked with anticipation at every tree, around each corner. Hoping, expecting to find something new, something never seen before – a 'novelty' to send back to those eagerly awaiting the results of each foray into the natural world. The excitement and adventure is conveyed as he describes to the eminent ornithologist, John Gould, his feelings regarding the *Rattlesnake*'s visit to New Guinea:

Our New Guinea prospects, however, are better, and with them are associated in my mind visionary prospects of Dendrolagi, Cucsi and Birds of Paradise, jumbled up with imaginary skirmishes with the natives.[5]

Lizard Island. (Oswald Brierly, watercolour, in 'Sketches on board the HMS *Rattlesnake*': Mitchell Library, State Library of New South Wales)

Jetty boat swamped when lowered to catch sea birds.

'Jetty boat swamped when lowered to catch sea birds.' (Owen Stanley,
watercolour: Mitchell Library, State Library of New South Wales)

MacGillivray was striving to paint a picture – for himself and for his colleagues in England. His letters, journals and publications are an attempt to capture his environment in every detail. His aim, to identify every possible inhabitant of each environment he encountered. He proudly informed Professor Edward Forbes that during one of the cruises of the *Rattlesnake* they had had the opportunity to land on 37 islands along the coast, 'several of which, from their small size, or the length of our stay, were thoroughly ransacked'.[6] Despite his choice of words, MacGillivray was far from indiscriminate scientifically. His strength was his ability to be selective: to identify what items would provoke the interest of his colleagues and patrons back in England, and to pursue them with determination.

MacGillivray's love of fieldwork is perhaps the area in which his father, William MacGillivray, Regius Professor of Natural History at Aberdeen, can be seen to have had the greatest influence upon him. William MacGillivray had strong views on those he branded 'cabinet collectors', who paid others to collect on their behalf and who would then spend their time describing and classifying without ever observing the species in the field. William's enthusiasm for immersion in fieldwork is seen in the introduction to his abridged version of William Withering's *A Systematic Arrangement of British Plants* (first published in 1830):

> Its object is to induce the young to betake themselves, when occasion offers, to the fields and woods, the mountains and shores, there to examine for themselves the rich profusion of nature.[7]

John MacGillivray 'betook' himself into fields, mountains and woods whenever the opportunity arose. Despite his occasional complaints that those benefiting from his exertions had little understanding of the work required, his writings display a real relish for adventure and living rough in the bush; for contact with the 'natives' and a constant hunt for the new.

He was also very proud of his abilities, and obviously enjoyed the rough spectacle he presented and the physical labour involved. In an 1849 letter to Edward Forbes he described his dress and kit as including:

> 'reef boots', flannel trousers, leather belt, check shirt open in front, a shooting jacket which sadly wanted mending and washing, and an old straw hat shading a sunburnt and unshaven visage. And then with the thermometer at 90° in the shade I had to carry water, ammunition, skinning materials, a double barrelled gun, insect net, collecting boxes, a quantity of Botanical paper and boards, besides two days' provisions.[8]

After keeping watch for hostile natives, rising early and staying up late to search for collections he exclaimed 'Yet what was the result of this fagging? A few birds, none of which were rare, about 25 species of plants, 3 or 4 insects and a Helix!'[9]

Coral Haven Canoe. June 19. 1849.

Figure head of Canoe — Coral Haven — Louisiade Archipelago

Figurehead of a canoe, Coral Haven, Louisiade Archipelago. (Owen Stanley,
watercolour: Mitchell Library, State Library of New South Wales)

The frustrations of a 'mere collector'

Unlike his father, MacGillivray failed to translate his fieldwork, observations and collecting into a successful scientific career, although William MacGillivray's work is certainly more appreciated now than it was during his lifetime. He too is reputed to have had a rather irascible and impetuous nature, and he made a number of powerful enemies throughout his career. MacGillivray senior was not a part of the powerful networks and scientific community that would have been useful in launching his son's career. In one telling description, Charles Darwin – despite professing enthusiasm for William MacGillivray's work on the birds of Scotland – observed that he 'had not much the appearance and manners of the gentleman'.[10]

An echo of this sentiment can be felt in Sir William Hooker's comments regarding John MacGillivray's *Narrative of the Voyage of HMS* Rattlesnake, which he felt was 'better than could be expected from a man of restless uncouth appearance and manners'.[11] Many of the observations made regarding the character of William MacGillivray, and the limitations that this character placed upon his work, could equally be applied to his son.

At the height of his career John MacGillivray was aware that he was not making the transition from collector to scientist, but he laid the blame of this failure upon the heavy workload of an official collector. In a letter to Edward Forbes he complained that 'having been unassisted to make collections in all the departments, my duties too often merge into those of a mere collector and preserver of specimens'.[12] This distinction would have been emphasised as he observed shipmate Thomas Huxley's deliberate, single-minded determination to forge a scientific career through his analysis and description of the Medusae family during his time on board the *Rattlesnake*. MacGillivray lacked Huxley's drive and vision in this respect. The time taken by fieldwork combined with the required preparation and packing of specimens – a task he complained to Hooker, took 'all day and half the night'[13] – left little room for the excursions of the mind required for minute description and analysis.

MacGillivray was, however, aware of the significance of his descriptive work. Just as eagerly as those back in England awaited the fruits of his fieldwork, he awaited their opinions in order to confirm that his 'novelties' were indeed such, that the judgement he had shown in acquiring his specimens was correct, and that the risks and deprivations he had endured had been worthwhile. To Forbes he wrote 'You must know how cheering it is to me so far from home as I am to hear of any novelties that may have been transmitted by me'.[14]

To another correspondent, Adam White, he wrote of the great encouragement that he gained from the news of the reception of his novelties.[15] Although he may never have succeeded in making a name for himself in scientific circles, he is remembered in the scientific names of a number of birds, shells and plants. In his natural history notebook – found in his rooms after his death – a small list of some of these appears on one of the pages. It was one way he could mark his own legacy to the world of natural science.

Fortress at Port Essington

The fortress at Port Essington. (Owen Stanley, watercolour: Mitchell
Library, State Library of New South Wales)

Dismissal and disgrace

MacGillivray's travels on board the Admiralty's surveying vessels are well documented. Harder to piece together, but no less fascinating, is the latter part of his career. The great shift in MacGillivray's fortunes occurred with his dramatic dismissal from HMS *Herald*.[16] The charges brought against him were grave: insubordination, intoxication, selling his collections for personal gain and other financially scandalous behaviour. The verdict of the Board of Enquiry was conclusive. On 26 April 1855, John MacGillivray was dismissed from service.

For his captain, Henry Mangles Denham, the relief at being rid of him was immense. He wrote that with MacGillivray's dismissal he felt 'as if a blister was removed from my heart'.[17] When Huxley heard of MacGillivray's difficulties he somewhat scathingly commented 'It is most lamentable that a man of so much ability should have so utterly damned himself as MacGillivray has, but he is hopelessly Celtic'.[18]

Life on the fringes

MacGillivray himself was silent regarding his dismissal. No record survives of his thoughts and feelings. He did not attempt to clear his name or justify his actions. No clear account exists either as to how the enquiry proceedings took place. The picture that emerges from the fragments that remain is of a painful, passionate, bitter and very public dispute as tensions, simmering for years in the close confines of the vessel, finally erupted. This enquiry would shape the rest of John MacGillivray's life. It would also colour the way in which his career, contributions and character would be regarded, recorded and related until the present day.

MacGillivray's situation as he stepped ashore in Sydney in April 1855 was serious. He was in utter disgrace. He had insulted and offended key figures both in Britain and in his new home. He had debts to pay, a family to support and no real prospect of being able to do either. Where he went and how he survived is not known. MacGillivray effectively disappeared into the city for over a year and was lost.

It was left to Huxley and Hooker to raise the funds required to send MacGillivray's wife, Williamina, and three children to Australia. Sadly, Williamina died on board the *Washington Irvine* two weeks out of Sydney. John Gray, Williamina's brother, wrote to Huxley thanking him for his kind attention to his sister and informing him of her death. He reported that the children were quite well and now in good hands. These hands, however, were not those of their father. Of him Gray reported that he was doing 'no good' and that any hope of reform was 'now doubtful'.[19]

Despite (or perhaps because of) the length of time MacGillivray had spent in and around Sydney during his career, he had failed to ingratiate himself with the dominant forces in the Sydney scientific community – most particularly the Macleay family. Here again Huxley had the advantage, having successfully cultivated a good relationship with

The *Herald* in Feejee [Fiji]. (James Wilson, hand-coloured lithograph: National Library of Australia)

William Sharp Macleay. Another Sydney scientific figure, Gerard Krefft, who beat MacGillivray to a curatorial position at the Australian Museum, wrote that:

> John MacGillivray, the well known naturalist of HMS 'Rattlesnake' was one of the competitors, and this gentleman would certainly have carried the day had not the Macleays hated him thoroughly because he was *a clever man*. I confess Mr MacGillivray (with all his failings) was superior to myself.[20]

Wanderings in the Pacific

Had MacGillivray successfully entered into Sydney's scientific circles or gained the patronage of figures such as the Macleays, he may have had more success in rebuilding a career in his newly adopted country. As it was, he was forced to seek his fortunes elsewhere, leaving Australia on 11 February 1858 on the appropriately named brig *Spec*, bound for New Caledonia.[21]

For the next two years MacGillivray worked as a private collector in New Caledonia,[22] and then Vanuatu, based for some time on the island of Aneityum. For a short time he worked for the sandalwood trader James Paddon and was employed by him to undertake

James Fowler Wilcox, one of MacGillivray's shipmates on board the *Rattlesnake* and later business partner in Grafton 1864–1866. (National Library of Australia)

a cruise of the Torres Strait and Cape York in search of sandalwood, beche-de-mer and other tradable commodities. MacGillivray wrote up this (spectacularly unsuccessful) cruise in a series of 11 articles for the *Sydney Morning Herald* in 1862 under the title of 'Wanderings in Tropical Australia'.[23]

Between 1864 and 1866 MacGillivray worked as a private collector in partnership with James Fowler Wilcox, based in South Grafton in northern New South Wales. (Wilcox too had served onboard HMS *Rattlesnake* as a private collector for Captain Owen Stanley's father, the Bishop of Norwich, on behalf of the Ipswich and Norwich museums.) The business relationship between the two men did not work as well as either of them had hoped and ended on an acrimonious note, with MacGillivray swearing to 'leave this house never to return'.[24] Despite this, MacGillivray obviously enjoyed his time in the area delivering a series of successful lectures at the Grafton School of Arts and playing a key role in organising material to be sent from Grafton for the 1866 Intercolonial Exhibition in Melbourne.

After dissolving the partnership, MacGillivray moved back to Sydney where he undertook 'daily conchological exercises' for James Cox, organising and assisting to catalogue his land shell collection. When he died, MacGillivray was busy planning an extended collecting trip to northern Australia.[25] Although excited at the prospect of the trip, he wrote that he must first complete his work for Cox: 'I shall do it, if the asthma doesn't choke me meanwhile, or one of the wild bulls of Bashan (or the county of Cumberland) stick his horns into my gizzard and bring my carcass before the Coroner'.[26] This was from the last known letter written by MacGillivray. He died on 6 June 1867, alone in his attic retreat.

MacGillivray was a natural communicator, perhaps some- what obstreperous, certainly proud, but blessed with some charm and charisma. Despite the disappearance of MacGil- livray Bluff, his name lives on in other forms. A rare petrel, a tropical reef and a number of shells and plants carry his name.[27] His narrative of the *Rattlesnake* remains one of the outstanding accounts of nineteenth-century voyaging in the Pacific, and a closer examination of his career and contribu-

tions will help to place him as a key figure in the development of the natural sciences in Australia.[28]

The obituary published in the *Clarence and Richmond Rivers Examiner* perhaps best captures the complexity of this remarkable man and is a reminder of the qualities that have since been forgotten:

This gentleman, lately a resident amongst us, died on Thursday last, in Sydney. It is to be feared that his death was accelerated by indulgence, the temptation to which he could not always resist, but in spite of this weakness, he was a man of honor and veracity, a firm friend and a most amiable companion. His acquirements as a scientific naturalist were of the highest order, and accompanied by a remarkable absence of pretence or self- sufficiency ... the minuteness and accuracy of his information on all subjects connected with his noble profession; his sly humor, always void of offence; his liberal spirit and strong sense made him a remarkable person. He had the cultivation of a scholar, and the heart of a gentleman. His one failing he shared with but too many: his merits were all his own.[29]

Rhynchotrochus macgillivrayi, first collected by MacGillivray on the voyage of HMS *Rattlesnake*, and described and named in MacGillivray's honour by Forbes in 1851. (Photo H Barlow: Australian Museum)

The oldest fish in the Australian Museum: an orangeband surgeonfish *Acanthurus olivaceus*, collected in 1858 in the New Hebrides (Vanuatu) by John MacGillivray. (© Australian Museum)

60. The wandering ALBATROSS in the INDIAN OCEAN. July 1842.

The wandering albatross in the Indian Ocean. (Edwin Augustus Porcher:
National Library of Australia)

indigenous encounters

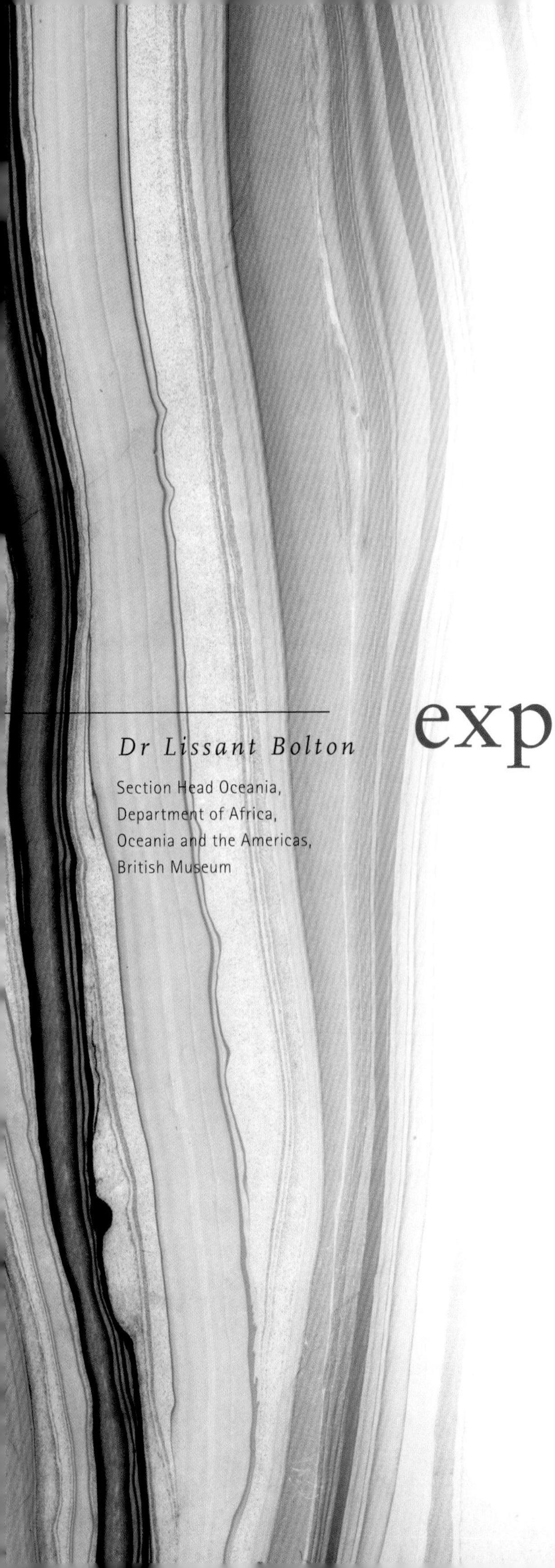

Dr Lissant Bolton

Section Head Oceania,
Department of Africa,
Oceania and the Americas,
British Museum

explorers and traders of South Papua

When European ships like the *Rattlesnake* and *Bramble* began exploring the western Pacific islands north of Australia in the late 1840s, they were, from their perspective, engaged on journeys of discovery in unknown lands. As they sailed, mapping coastlines, collecting scientific specimens and artefacts, and drawing and describing what they saw, they frequently encountered canoes. The western Pacific waters were alive with canoes.

People there were going about their business as they had for millennia. Small canoes plied back and forth across lagoons and along rivers, and people made regular, even daily journeys between offshore islands and the adjacent mainland to work in gardens or to visit kin. In some places there were war canoes which frequently set

Canoe prow ornaments from the Louisiade Archipelago, Papua.
(Collected by Owen Stanley: © The Trustees of the British Museum)

A Rossel Island canoe, drawn from above –
perhaps from the deck of HMS *Rattlesnake*.
(Oswald Brierly, sketch, 1849: Mitchell Library,
State Library of New South Wales)

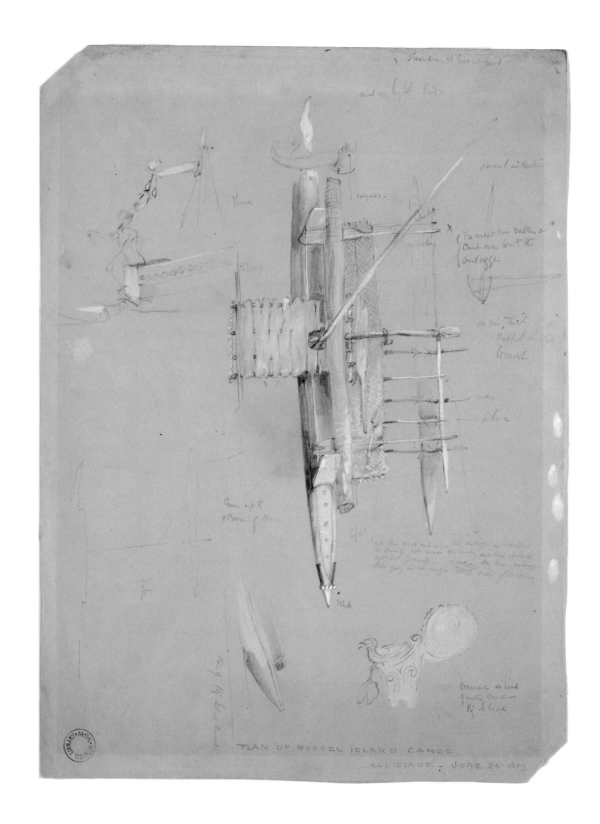

off on raids. Large trading canoes made long journeys, often along regular trade routes, following winds and currents well-known to those who directed these expeditions. The trading canoes were mostly sailing canoes, sometimes over 14 metres long, on which ten or twenty people might live for days at a time. They were trading goods from one place to the other, sometimes trading for goods for their own communities, and sometimes operating as middlemen or merchants, taking goods from one place to another, in exchange for something else again. So when European explorers encountered the indigenous peoples of this region, they were not the only mariners traversing foreign waters, nor the only visitors who the locals received.

Discovering the discoverers

In Captain Moresby's account of the 1873 voyage of HMS *Basilisk*, which mapped the south-eastern coast of Papua New Guinea, he reports a series of such meetings. When the *Basilisk* was moored in Fortescue Strait, near Basilisk Island, the ship was visited 'by some trading canoes of large size which came from the east, we supposed from some of the Louisiade group' – that is, some 200 kilometres away.[1] In order to better survey the maze of reefs and small islands in this area, Moresby decided to make an exploring expedition in the ships' boats over several days. While on this expedition, they ran out of fresh water and were not able to find any on shore. Moresby continues:

seeing one of the large trading canoes standing in for China Straits, we gave chase, to the great alarm of its crew, who numbered about fifteen and had several women and children on board. There was no wind, so we soon came aside, and when the astonished creatures found we meant them no harm, they gladly supplied us with water from coconuts, the orifices of which were stopped with grass, and pointing to a large village in China Straits, made signs that we could obtain plenty there. Accordingly, we pulled in for the western shore of this third new island, named by us after the senior lieutenant, 'Hayter,' by the discovery of which we had now cut off in all forty miles from the supposed length of New Guinea. As we approached the village, which was situated partly on a small islet, and partly on the mainland, to which it was joined by a reef, numbers of canoes came out to meet us; manifesting some doubtfulness, until they had communicated with their friends in the large trading canoe, after which they became assured, and crowded round us.[2]

Moresby's sense of 'discovery' was powerful. He also marvelled at seeing and mapping an area that 'had never previously been visited, and was actually unknown as to its conformation (as far as I have been able to discover any record)'. Yet he does not measure his 'new' knowledge against that of the people who already lived there, for of course these islands, their coastlines, currents, natural history and social arrangements were well known not only to its residents, but also to those who voyaged around the region in the trading canoes – like the group who directed Moresby and his men to a source of fresh water

A CHINA STRAITS CANOE

A small local sailing canoe from China Straits. (Anon, etching, in
J Chalmers, *Pioneer Life and Work in New Guinea*, 1895: Mitchell Library,
State Library of New South Wales)

and introduced them to the villagers who controlled it. The fact that these areas were well known to others does not diminish the achievements of these early European voyages of exploration and survey, but it provides a valuable counterpoint to our understanding of their achievements. To put it another way, their achievements direct our attention to the achievements of those who had explored, navigated and settled these islands before them, and whose trade in goods and knowledge both preceded and was contemporary to those early European voyages.

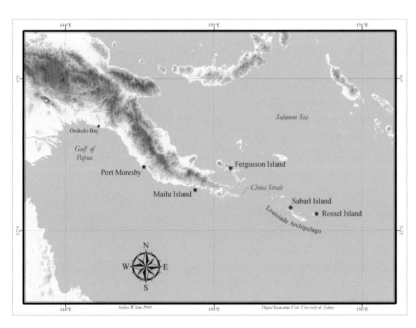

Papua New Guinea. (Andrew Wilson: Digital Innovation Unit, University of Sydney)

The settlement of south-eastern New Guinea

The island of New Guinea has been continuously inhabited for over 40 000 years. When it was first settled, it was not an island at all, but was joined to Australia. Sea-level rise separated the island from the mainland of Australia about 8000 years ago. Archaeological evidence shows that at least a millennium before that separation, about 9000 years ago, people were practising agriculture in the island's central highlands – an innovation that predates the development of agriculture in the Middle East. Most of the island of New Guinea is now inhabited by people speaking languages belonging to a distinctive language group – Papuan languages – descended from those ancient inhabitants.

Along the coast of New Guinea and on the offshore islands, however, people speak languages belonging to another family – Austronesian. Austronesian-speakers began moving into the western Pacific about 3500 years ago, eventually sailing their canoes right across the Pacific to Easter Island and New Zealand, and settling all the islands along the way. They reached New Zealand about 1200 years ago. This history of exploration, colonisation and settlement may have begun further west, but it has been traced backwards as far as the islands north of New Guinea – the Bismarck Archipelago.

The settlement of the western Pacific by these Austronesian-speakers is recognised archaeologically by their pottery, which bears distinctive incised and stamped designs on the outer surface. Both the pottery and its makers are known as Lapita. It is found through island Melanesia (the Solomon Islands, Vanuatu, New Caledonia) and further east again in Fiji and Tonga – although not in Polynesia or the islands of the eastern Pacific – in sites dating up to about 2500 years ago. After that it seems to fall out of fashion, and other pottery styles emerged in the areas the Lapita people had settled.

Because the Lapita settled the whole Pacific region, theirs is the language from which all Pacific languages developed. Linguists comparing these have identified some words common to all of them. These words probably existed in that original language in the same way that many European languages share words drawn from Latin. By identifying those common words it is possible to suggest some characteristics of Lapita society which cannot be traced in the archaeological record. They include words for sail, steering oar, bailer and anchor, plus words for making a sea voyage, for cargo and for boat owner.[3] This suggests that they had a sophisticated sea-going culture. The Lapita people and their descendants were great explorers, sailing not just to places visible on the horizon, but discovering and travelling back and forth from islands far beyond it. Like European explorers, they would have given names to the places they discovered and taken back reports to their homelands of new islands in which they could settle, and the resources to be found there.

Essentially sea-going people, the Lapita people did not venture far into the mainland of New Guinea, and if they explored its southern coast, they did not settle there. It seems that until about 2000 years ago, the south coast of Papua was inhabited by a small number of Papuan-speaking hunter-gatherer groups, moving through the landscape. Then about 2000 years ago, these coastal areas – the region later explored for Europe by the *Rattlesnake*, *Basilisk* and others – were settled by Austronesian-speaking colonists who arrived by sea and settled first on offshore islands, beaches and headlands. These were not the Lapita people, but were a group of their descendants. All along the south Papuan coast, their settlements are characterised by another distinctive style of pottery, which archaeologists call Early Papuan Pottery, suggesting that the same people settled the whole coast.[4] Over time, these settlers again developed more localised pottery styles, or later gave up making pottery altogether.

When these people first settled the south Papuan coast, they were in contact with both each other and with places further afield. Imported goods like obsidian (volcanic glass) from Fergusson Island can be found in their early settlement sites. These trade connections seem to have changed over time. Groups along the south coast today recall that their ancestors came from elsewhere, by sea: they recognise themselves as the descendants of immigrants. Neither archaeology nor linguistics can give details to this history, but surely these immigrants arrived because someone first explored along this coast and judged it suitable for settlement? The Austronesian settlers of the south Papuan coast were, like their Lapita antecedents, both explorers and traders.

In the centuries before European colonisation, the many small communities in this region were commonly in a state of hostility with their neighbours, and it was not safe to travel out of one's own area by foot. The rivers and the sea provided a safe way to travel. As well as voyages of exploration and trade, some communities also mounted regular war expeditions, mostly attacking known enemies in adjacent regions. There were war canoes along the south Papuan coast, but less is now known about them. Most records focus on the trading canoes that voyaged in overlapping regions, ultimately linking the whole coast of the island in a chain of interconnection.

The Motu traders of Port Moresby

The trading canoes which Moresby encountered were some of the many thousands of canoes operating in the waters of the region. Some of these trade networks have been documented in detail and have become famous – most notably the Massim *kula* trade network linking the islands of south-east Papua, which was described by Bronislaw Malinowski[5] – but there were many others. At the end of nineteenth century, two groups dominated the trade networks along the south Papuan coast: the Mailu and the Motu, and archaeology suggests that both had maintained this dominance for the previous thousand years.[6] The Mailu, who lived on an island off the east coast, were merchant traders, acting as middlemen in a series of annual, seasonal exchanges. The Motu, on the other hand, lived near what is now Port Moresby, and from there made an annual trading voyage to the Gulf of Papua –

a round trip of about 800 kilometres – exchanging pots and valuables made from shells for sago and canoe hulls. In 1910, FR Barton evocatively described the annual Motu trading voyage:

> Every year, at the end of September, or the beginning of October, the season of the south-east trade wind being then near its close, a fleet of large sailing canoes leaves Port Moresby and the neighbouring villages of the Motu tribe on a voyage to the deltas of the Papuan Gulf. The canoes are laden with earthenware pots of various shapes and sizes which are carefully packed for the voyage in dry banana leaves. In addition to these, certain other articles highly valued as ornaments (and latterly foreign made articles of utility) are also taken for barter. The canoes return during the north-west monsoon after an absence of about three months, laden with sago which the voyagers have obtained in exchange for their pots and other articles.[7]

The trading canoes, called *lakatoi*, were re-constructed every year for the voyage. In April or May, individual men decided to make the voyage and recruited their crew, and in August they began overhauling and caulking the large dugout hulls, then lashed them together and built a platform or deck above them. The masts were stepped, and the sails manufactured. The sails were of plaited mats sewn together to form the 'crab-claw' shape which made the *lakatoi* so visually arresting. Once the *lakatoi* were ready, they held races across the harbour. Groups of young women collected on the projecting platform at the prow of each canoe and, as Barton described it 'danced there with great vigor, the springy nature

A *lakatoi* with its sails lowered, probably near Port Moresby, 1903–04. (Cooke Daniels Expedition collection © The Trustees of the British Museum)

A fleet of *lakatoi* starting for the Gulf of Papua. (Anon, etching, from J Chalmers, *Pioneer Life and Work in New Guinea*, 1895: Mitchell Library, State Library of New South Wales)

of the platform adding largely to their lively movements'.[8]

In October 1883, the missionary James Chalmers joined a *lakatoi* on its trading voyage west, and kept a diary which was later published in 1895. He described the winds and currents, seeing the lights of other *lakatoi* in the distance at night, and the anxieties of the crew when, nearing their destination, they had to guide the *lakatoi* safely through the surf and up the river to the village. He also related that a large fighting canoe came out to them from a place he called Maclatchie Point, wanting to appropriate the *lakatoi* and its contents – a possible conflict which, he reported, he averted himself. Up to 30 people travelled on each *lakatoi*, and Barton calculated that 20 *lakatoi* made the journey in 1903, taking in total nearly 26 000 pots to their trading partners.

Once the *lakatoi* reached their destination, the cargo was traded, and the crew settled down, living on the *laka-*

FLEET OF LAKATOIS STARTING FOR THE WEST

toi, to wait for the purchasers to prepare the sago they had promised as payment. Some crewmen also traded shell arm-bands in exchange for logs which they hollowed out on the beach into dug-out hulls. They then dismantled their *laka-toi* and reassembled it to include the new hull, thus taking it home.

Girls dancing on the prow of a *lakatoi*, probably near Port Moresby. (RA Goodyear: © The Trustees of the British Museum)

The Mailu and middlemen trading

Mailu Island, further east along the south Papuan coast from the Motu, provided only a limited area for agriculture. As a result, the Mailu Islanders were dependant on trade. Like the Motu, they built large double-hulled canoes, called *oro'u*, which also used crab-claw sails woven from pandanus. The Mailu, however, did not make just one voyage each year, but a whole series, in each case trading different goods to different trading partners in an annual cycle which began in July or August and continued until late January. Malinowski described the sequence of trade in detail:

> The trading was essentially seasonal and regular, each expedition forming a step in a consecutive series of ceremonial transactions and industrial activities (making of armshells, sago, pottery ...) and everything leading up to the final expedition which brought back the all important pig supply.[9]

The trading season was followed by a feasting season which was of great importance in the life of the community. By this regular pattern of trading, the Mailu linked a number of different groups along the south Papuan coastline.

Traders never just traded in goods. They also traded in information, in stories and gossip. They learned about the whole region that they visited, and became familiar with the knowledge and practices of its peoples. As Frank Teisler comments about another trading group on the north coast of Papua New Guinea (the Murik), the paradox of these outrigger canoe peoples was that they were the ones who acquired prestige in the region, not the groups upon whom they were dependant.[10] The Mailu were regionally important despite the poverty of their land, and it is likely that, like the Murik, part of their importance was that they knew a great deal about the culture and politics of the whole area they visited.

The canoe voyagers of this region also had an extensive knowledge of their sailing environment. They knew about

Pottery being prepared to send to the Gulf of Papua on a *lakatoi*, Hanuabada
village, near Port Moresby. (WG Lawes: © The Trustees of the British Museum)

the sea – its tides and currents, the wind and weather. So crucial were the winds that both the Mailu and the Motu named the seasons of the year after the prevailing winds of each. Lessons in seamanship were learned in youth. Mailu boys learnt to sail by playing with canoes. Models of the large crab-claw sail double canoes were made for them, and boys spent whole days following their canoes as they sailed across the shallow water of the bay in front of the main Mailu village. Boys also had small out-rigger canoes – too small for a grown man – in which they sailed near the village. Malinowski comments that sometimes they ventured farther out, even in fairly rough weather, with amazing skill and daring.[11]

The Mailu trade routes connected to those both to the east and west of their region. There were, in particular, strong connections between the Mailu district and the major trading communities around the eastern end of Papua, by which objects were passed from group to group over long distances. News of far-off places would have passed with these goods. Indeed, as Malinowski observed, ideas and customs may have travelled from the mouth of the Fly River in the far west of Papua, and beyond, as far east as Woodlark and the Trobriand Islands and the north-eastern coast – although he suggested that the main region of interconnection was from Mailu to the east.[12]

The history of the original indigenous exploration and settlement of what is now Papua New Guinea is still being uncovered. Archaeological discoveries often confirm details evi-

dent in local histories and myths in individual communities, or make sense of histories that earlier researchers could not interpret. What is clear is that the original explorers of this coastline, and the settlers of it, were as intrepid adventurers and as skilled seamen as the Europeans who came after them. Long before the Europeans started building a network of explorers and navigators who collected knowledge and artefacts from this corner of the world, its indigenous peoples had built and maintained their own equally sophisticated network of knowledge trade and exchange.

The Mailu trade network was, however, greatly diminished over the period of the twentieth century. Indeed, many of the coastal trading networks are now much reduced or have even disappeared, partly as roads and airstrips have been built on the mainland. Neither Motu nor Mailu canoes now sail on trading expeditions in the way they once did, but the traditions of sea-going trade remain stronger in the offshore islands and archipelagos, for example between the islands beyond the eastern point of Papua New Guinea. In Sabarl, an island of the Louisiade group – near where the *Rattlesnake* voyagers collected canoe ornaments and Brierly sketched canoes from the ship – people remained very dependant on their vessels. Debbora Battagalia reports a young married Sabarl man saying as recently as the late 1970s:

When you lie awake at night and you can't sleep, you are
worrying. Always you are worrying about only two things:
canoes and [food] gardens. It is always these: canoes and
gardens.[13]

Mailu men working on their canoes, south coast Papua. (Frank Hurley, *Photograph Album of Papua and the Torres Strait*, 1921: National Library of Australia)

Plaited pandanus sail laid out in the main street of Mailu village. (Frank
Hurley, *Photograph Album of Papua and the Torres Strait*, 1921: National
Library of Australia)

Mailu Village, Mailu Island, south coast Papua, where boys learned to sail their small canoes. (Frank Hurley, *Photograph Album of Papua and the Torres Strait*, 1921: National Library of Australia)

Dr Jude Philp

Macleay Museum, University
of Sydney

days of desolation on the New Guinea coast

HMS *RATTLESNAKE*

When Thomas Huxley returned to Australia for the last time, frustrated with his lack of opportunity for exploration, weary of cramped ship conditions, bored with everything and yearning for his fiancée, he brought with him 'some combs made of tortoise shell I got in New Guinea. There's one for you and one for Mrs Griffiths if she likes to have them – I meant to have some more made – but forgot – then again among this miscellaneous collation of articles you will find a letter for you which was written in the days of desolation on the New Guinea coast'.[1] This is one of the few written statements surviving from the voyage of HMS *Rattlesnake* that give some indication of why the Europeans were interested in acquiring objects from the Papuan peoples they encountered.

That New Guinea held the true excitement of the voyage for the crew and officers of the *Rattlesnake* is obvious from the collections given to the British Museum and the writings of three of the 180 on board: Thomas Huxley, assistant surgeon, John MacGillivray, naturalist, and Charles Card, clerk. Of greatly different temperaments, these three men depicted their encounters with local peoples in ways that also allow us to learn something about what local people made of this strange company of ships. The exotic tokens of the four-year

expedition that were traded by officers and crew alike were later traded on, sold and given away to friends and family. Just under 250 were donated to the British Museum – becoming official markers of the *Rattlesnake*'s journey. From the writings of Huxley, MacGillivray and Card it would seem that officers had the most contact on land with locals and therefore a greater opportunity to point to things that they were interested in acquiring, but that the greatest opportunities for trading happened on board amidst the more egalitarian company of crew and officers. Through this collection it is possible to see what the local peoples willingly traded, and also what the British officers desired and thought of as representative of their voyage.

'Dawdling about ... on the coast of New Guinea'[2]

This voyage of HMS *Rattlesnake*, by then a 24-year-old three-masted frigate, was captained by Owen Stanley. It would be the last voyage of this experienced navigator and surveyor, as he died on the ship's return to Sydney at the beginning of 1850. The accompanying schooner *Bramble* was captained by Lieutenant Charles Yule, one of a number of men (including MacGillivray) who had spent 1845 in the same waters on the surveying voyage of HMS *Fly*. Because the *Rattlesnake*'s survey principally involved perfecting and filling in details missing from such previous surveys, their work was particularly tedious and painstaking, often involving the ship stand-

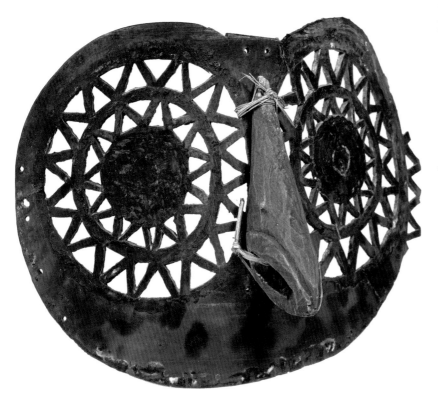

Turtleshell and wood mask collected from Naghir (Mount Ernest) Island in 1849. (© The Trustees of the British Museum)

Necklace acquired at Erub (Darnley) Island in the Torres Strait, 1849. (© The Trustees of the British Museum)

ing uncomfortably at sea, or anchored for days on end, while the *Bramble* and the accompanying pinnaces and cutter took detailed bearings of the coastline and soundings from the ocean floor.[3]

Between June 1849 and January 1850 the ships surveyed the New Guinea coastline and made further soundings in Torres Strait, and it is solely during this period that the 248 objects later donated to the British Museum were collected.[4] It is intriguing, and surprising, that only eight of these were the cultural products of the Aboriginal peoples of Australia, all from Cape York. One of the few explanations for this is that interest in Britain for Aboriginal artefacts had waned. Instead, for the Europeans on the *Rattlesnake* voyage, it was the novelty of products from New Guinea – heralded as the first collections from these 'new' peoples – that was their principal significance.

While theirs may have been the 'first contact' with some local peoples, these British visitors had some guides to their visit into New Guinea. Some of the crew already had previous experience from earlier survey expeditions, which also yielded a variety of written accounts – which MacGillivray painstakingly outlined over twelve pages in the official narrative.[5] It is apparent that the local people also had their sources of information concerning the visitors, as can be seen in the trading encounters described in the various accounts. On approaching the Louisiade Archipelago, for example, locals held up un-worked turtle shell plates to trade – an action that, since it was not a commodity of indigenous trade, implies some prior knowledge of European desiderata.[6]

The complexities
of trade

The islander people encountered by the *Rattlesnake*'s crew are representative of one of the most culturally diverse areas in the world. From the Torres Strait to the Louisiade Archipelago and along the Papuan coast to Redscar Bay and beyond, there occur dramatic differences in languages, customs, dress and technology. As Huxley wrote at Rossel Island, in disbelief at the varieties of native craft, 'could there possibly be a completely different kind of canoe from people just 20 miles distant?'[7]

While diverse, these people were nevertheless united through broadly similar ways of trading. Shell goods, agricultural produce, minerals, medicine and animal products circulated between peoples across the Arafura sea, into the landmass of Papua New Guinea, and beyond. Often characterised as gift exchange, these distinctive politico-economic systems typically involved a lifetime of encounters where gifts were given and complex negotiations were entered into to determine the appropriate equivalent return gift. Gift exchange was typified by the strong social and familial bonds that unite a person and their trade partner, and also by the way that success in trade marked the success of the individual in other aspects of their life. At most points what was valued, how trade was negotiated and how trading partnerships were developed and maintained, differed from culture to culture. Many exchange systems also included opportuni-

Trading with the Natives at
Coral Haven June 1849

HMS Rattlesnake bartering with native Canoes off the Pig Island
Louisiade Archipelago — June 1849

Owen Stanley's vision of the people of the Louisiade Archipelago trading with the
Rattlesnake officers and crew, June 1848. (Owen Stanley, watercolour: Mitchell Library,
State Library of New South Wales)

ties for bartering and smaller side trades. In various cases, gift exchange of objects and foodstuffs was the most material aspect of a more complex engagement which also included esoteric knowledge, songs and dances, and the great danger associated with the voyage across the reef-studded seas and into foreign territories. The tense negotiations of indigenous trading excursions could also become violent. It was these established trade networks which the *Rattlesnake*'s voyagers at times intersected, and these local cultural practices which the ship's crew unwittingly mimicked.

One such occasion of Europeans conforming to local trade patterns occurred at the Louisiade Archipelago when, like trading partners from distant islands, the *Rattlesnake* arrived during harvest time and requested yams and small shell valuables, offering cloth and, significantly, axes in return. 'Axes' noted MacGillivray, 'were more prized than any other article'.[8] While the blades offered by the British were iron, blades used by the Vanatinai (Sudest) Islanders were made of nephrite. But these were not made locally, and like iron had been traded from far distant places.[9] This is not to say of course that the British understood their trade as falling into local patterns. Indeed at one point MacGillivray wrote that one recipient of an axe 'was quite speechless – not understanding the nature of a gift'.[10] It must have soon been apparent to the locals that the British did not value their axes very highly, with one axe offered for a quantity of yams, one for time given, and one for a wooden bowl.[11]

Trade in the Louisiade Archipelago was principally conducted on the ships and even as the boats moved about, Mac-

Headdress from Moa (Banks) Island, collected 1849. Cassowary feathers were a common trade item between Islanders, Aboriginal people and the mainland Papuans. (© The Trustees of the British Museum)

Gillivray and Huxley recognised particular people from one day to the next, which meant that the local people could establish the particular wants of these new trading partners. One Sunday the crew's interest was raised when they sighted, but were unable to trade for, a human jaw armband worn by a man in a visiting canoe. The following day, Huxley wrote, 'several canoes came off this morning; one of them brought the [canoe] figure-head which was so much wanted yesterday, and bartered it immediately' and with it a 'jaw bracelet' and a nephrite hatchet.[12] Four of these jaw bracelets, along with shell neck-laces and ornaments, fishing gear, carved lime sticks and other paraphernalia attached to the pastime of chew-ing betel nut, pearlshell knives, combs and some baskets in the British Museum are the relics of these times.

Friendship and friction

In 1848 and 1849 the officers and crew spent long periods at Cape York. Here, interaction with Kaurareg peoples of Torres Strait and their Cape York neighbours was extensive, partly due to the amount of contact these peoples had had previously with foreigners passing through the strait, partly because of the amount of time the *Rattlesnake* spent there, and partly because in 1849 they had access to a bilingual speaker in the form of the shipwreck survivor Barbara Thomson who had lived with the Kaurareg for five years. Indeed throughout the Torres Strait, the ships were greeted by men and women who recognised MacGillivray or others by name from the visit of the *Fly* just three years earlier.

Turtleshell ornament associated with lime, collected from the Louisiade Archipelago in 1849. (© The Trustees of the British Museum)

Huxley and others exploited the personal relationships into which they were drawn: 'I established an especial friendship with Do-outou, with whom according to custom I changed names. He promised me a "coskeer" (wife) and all sorts of fine things when I came ashore.' What Huxley did not at first realise was that such relationships, where trade partners exchange names, were not just a token of friendship but also involved obligations, and the following day he not only met the 'good-looking' Kaeta but was surrounded by her relatives – who likewise requested gifts due from the bond of friendship.[13]

For the captain, even such light-hearted and friendly visits held the promise of treachery. Exceedingly cautious throughout the survey, seldom landing or allowing others to land, Stanley frustrated all who were excited at the prospect of meeting new peoples and alleviating the boredom and misery of life on board the ship. 'Today for the first time we have seen the coast of New Guinea' wrote Huxley on 12 August as the *Rattlesnake* prepared for what would be a 12-day-stay off the small island of Brumer off the coast. Here Stanley's caution tested all, with only two two-hour visits on shore allowed in the whole period.[14] For this reason it was often the local peoples who made the first move in contact, sailing or paddling out to the boats wherever they were moored.

Due to Stanley's caution, and the slow and methodical nature of survey work, local people had ample opportunity to view the foreigners for days at a time. That the 180 men and boys associated with the *Rattlesnake* regarded themselves as one group is obvious from the writings. Whether those on shore also recognised the *Bramble*, *Rattlesnake*, their pinnaces, the *Asp* and smaller boats as one unit is less clear. One example from their visit to Brumer Island is illustrative of the difficulties of understanding and intent. For ten days the crew enjoyed the music and dances brought by the locals, along with the enormous opportunities for trade, amusement and fresh food, courtesy of the islanders and mainlanders who came in great numbers to visit the foreign ships. The locals were in turn treated to costume dress-ups, fire-work displays, exotic food (the inevitable biscuit), drink, magic lantern shows, military drills and British popular dances. Such was the bonhomie between them that women as much as men came in canoes to trade and to take in the sights.[15]

The variety and extent of the collections made on these occasions is evidence too of their good relations. Women's skirts, bark cloth, men's feathered head ornaments, shell body ornaments, musical instruments, canoe decorations, fishing gear and even a paint pot are part of the 67 items from Brumer Islanders and mainland visitors. Just

Men encountered by the Europeans commonly carried arrows, spears and other implements for hunting, fishing and protection such as these collected from Redscar Bay in September 1849. (© The Trustees of the British Museum)

as dances were performed and exchanged, so too it would seem were the costumes worn, with articles of dress given out as presents. As Card described:

> The ladies then, after being rigged out in white petticoats and old shirts made into petticoats and several bead dresses, began their dance which was much quieter and steadier than the men's, as they merely joined hands and paced up and down the quarter deck keeping time with their feet; one or two were really good looking and were beautiful figures; they remained on board nearly two hours after which, being presented with a piece of iron hoop all round, they went down to their catamarans and managed to get Smith, one of our second class boys, down there and kissed him and hugged him all round, putting their arms round his neck and appearing greatly delighted; the boy too was just as well pleased as they were and kissed and caressed them in return

Pan pipes probably acquired after a performance by the Dufaure Islanders in August 1849. (© The Trustees of the British Museum)

Bamboo comb from Erub (Darnley) island: one of many objects collected from Torres Strait and Australia during the surveying voyage of HMS *Fly* between 1842 and 1846. (© The Trustees of the British Museum)

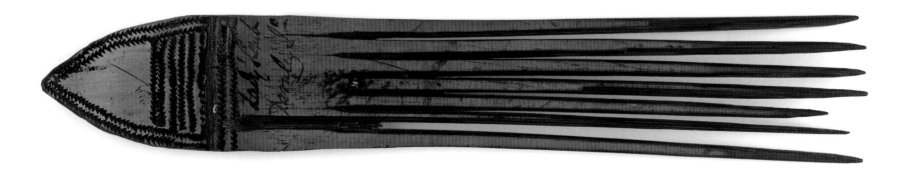

Bamboo lime container, possibly used for decorating dancers, collected from the Brumer Islanders in August 1849. (© The Trustees of the British Museum)

at which they were highly delighted and screamed aloud for joy and would hardly let him come up the side again.[16]

The *Rattlesnake* and *Bramble* left Brumer Island on the 29 August heading west separately on further surveys. On the 31st Yule, in the *Bramble*, became nervous as reportedly 60 canoes surrounded the ship, although he avoided any confrontation by moving swiftly off. The following day another canoe approached, this time with a man recognised from Brumer Island, who requested to board, showing friendly intentions. When Yule saw three more canoes apparently joining the first he, as Huxley put it, 'became excited and called for his gun, which was loaded with small shot'.[17] The rapidly paddling canoe parties were fired on four times. Huxley and Card shared their disgust at the actions in their journals. Card wrote:

> Yule thinks he has done something very brave and says he thinks they have got a pretty good lesson, while it is the opinion of nearly every one on board that it [was a] great piece of treachery on the part of old Yule and that he

deserves to have a couple of spears through him the first time he lands. He is so changeable too as he had no sooner got here than he gave an axe to the first native that came alongside so as to make chums of them and have been very friendly with them ever since they have been here.[18]

In the official narrative, however, MacGillivray remained silent.

'Far more civil than we imagined'[19]

As such confrontation suggests, the material documenting the culture of the people of New Guinea from the *Rattlesnake* is a collection born from a series of fragile relationships. Only with the accounts of meeting Kaurareg and the Brumer Islanders does there appear to have been a friendly and relaxed relationship sustained over some period of time. Yet there is a disparity in what went into the prestigious British Museum. A large number of Brumer Islander objects were collected, but seemingly nothing from the Kaurareg peoples of the Torres Strait.

This bias is also reflected in the composition of the collection that was donated to the British Museum, and in the selection of images to furnish the Admiralty-sanctioned official narrative. Perhaps this was because Jukes had done such a thorough a job documenting the Torres Strait and Cape York peoples in the *Fly*'s narrative, which was illustrated by professional artist Harden Melville's precise drawings and documented through the extensive collections deposited in the British Museum.[20] It may simply be that images, objects and stories about Australian Aboriginal and Torres Strait peoples were simply thought to be 'old hat'.

It is certainly true that the official *Rattlesnake* narrative privileges novelty and 'first encounter', demonstrating the place of the *Rattlesnake* amongst the great voyage narratives of Britain. Stanley's cautiousness in establishing friendly communications frustrated the possibility of insightful, systematic observation of mainland Papuans at every turn.

On dry land, however, the local peoples whom Stanley sailed past without interaction gained abundant information to be traded and gossiped about across the coast and with trading partners in far distant islands and into the hinterlands. Drills in the morning, practice with the guns, bathing on 'deserted' islands, fireworks, songs, the slow

This kind of pearlshell breast ornament, *mai*, is common throughout the Torres Strait. This one was collected from Moa (Banks) Islanders encountered on a neighbouring island. (© The Trustees of the British Museum)

A NATIVE DANCE AT DARNLEY I?

'A Native dance at Darnley': a typically descriptive image by the artist on HMS
Fly, Harden Melville. (From his *Sketches in Australia and the Adjacent Islands*,
1849: Mitchell Library, State Library of New South Wales)

Natives of Redscar Bay.

and odd movements of the ships' surveys – all these offered local observers a chance to understand more about the foreign people who were increasingly becoming a part of their world. As the Naghiri Islanders of Torres Strait sang in the 1840s:

Choki eenow good	[Tobacco no good
marki an gool kibou pah	white man ship dance and sing
sagoob pah sagoob bissikari nipa	tobacco biscuit knife
lakakinya uria nipa mangeeb.	they have gone to get for us.[21]]

This is the first known published image of the peoples of Redscar Bay and is one of many of TH Huxley's drawings included in the official narrative of the survey. (From John MacGillivray, *Narrative of the Voyage of HMS* Rattlesnake, 1852: Rare Book & Special Collections Library, University of Sydney)

A shell necklace typical of nineteenth-century coastal Mekeo and Motu people of the Redscar Bay area, and thus a novelty to the collections deposited in the British Museum. (© The Trustees of the British Museum)

relying on the locals

ALFRED WALLACE AND INDIGENOUS SAILING CRAFT

Professor Iain McCalman

University of Sydney

Alfred Russel Wallace was never a good sailor. Like his future friend Charles Darwin, he inclined to seasickness. On the other hand, he loved rivers. As a boy, his house in the Welsh border town of Usk had backed onto a fast-flowing river where he had watched in fascination as locals fished from coracles made of wood and skin. Working along the Amazon from 1854, he had also depended on a variety of locally built and navigated craft to carry him along some of the least-explored of that river's tributaries. He had learned to stow himself and his specimens in a succession of small native canoes called *montarias* that stank of rotten fish and animal hides. Occasionally he was forced to rely on even smaller and cruder dug-outs called *obas* that were hewn from a single log. Shooting down rapids in the Rio Negro under the guidance of Indian canoeists had not troubled him, but he always felt far less confident sailing in British ships on the open sea.

That Wallace came to believe in sailing with indigenous people, rather than on the larger vessels of his countrymen, symbolises his approach to naturalist exploration. His reliance on 'natives' – not least his Malay assistants Ali and Baderoon – and their technologies, stands in contrast to the methods of those earlier naturalists employed on British naval survey ships. It influenced and shaped some of Wallace's most important bio-geographical and evolutionary discoveries.

The ship from hell

AN INDIAN VILLAGE ON THE RIO NEGRO.

An Indian village on the Rio Negro. (AR Wallace, *Travels on the Amazon and Rio Negro*)

The river Usk and the town where Wallace spent the first five years of his childhood. (Photo Kim McKenzie)

Wallace's worst fears about British ships were realised in August 1852 on a return voyage from the port of Para, at the mouth of the Amazon, to Deal in England. He was travelling on a 235-ton brig, the *Helen*, with a cargo which included 10 000 bird skins, a large herbarium of dried plants, numbers of birds eggs, a small menagerie of live animals, and his books of sketches and notes. As a 31-year-old former trades-

man and self-taught scientist who depended on selling zoo-logical specimens to museums and naturalists, Wallace's whole future was tied up in this three years worth of specimens he had painstakingly gathered in the Amazon Basin.

He was reading in his cabin when Captain Turner popped his head through the door to say, 'I'm afraid the ship's on fire; come and see what you think of it'. Wallace was feeling decidedly unwell at the time. Seasickness had merged with a recurrent dose of malaria that had struck him two days after leaving port. Doses of calomel had temporarily stunned the fever, but he still struggled when clambering up to the foredeck to see 'dense vapoury smoke' billowing out of the forecastle.

Perhaps Wallace's debilitated state prevented this normally most practical of men from advising against Captain Turner's next move. Turner ordered his crew to open the forward hatchway to pour water into the hold. By letting in 'an abundance of air' – Wallace recorded years later – the captain transformed the smouldering heat into a leaping fire.[1] If he had caulked every crack and sealed the hatches with tarpaulins, he would have cut off the oxygen supply and smothered the flames. In retrospect, too, Wallace learnt that Turner ought to have packed wet sand around the 20 casks of a highly flammable natural lacquer *basalm of capivi* which they were carrying in the hold. Not realising that the liquid was so volatile, the captain had used wheat chaff as a packing substitute when he ran out of sand. Wallace could now hear the lacquer 'bubbling like some great cauldron' under the deck.[2] The old ship was as dry as a tinderbox, and the rest of the cargo, which included 120 tons of India rubber, was quick to ignite.

Dazed and overcome with 'a kind of apathy', Wallace shuffled down to his cabin, which was already smoke-filled and 'suffocatingly hot'. He grabbed a couple of shirts and a tin box containing a few old notebooks and some drawings, leaving behind a large portfolio of notes and sketches to keep company with the crates of specimens burning in the hold. On deck, meantime, the crew scurried to make the longboat and the gig seaworthy, tossing in sails, cordage, charts, and barrels of food and water. Meanwhile the flames ate through the skylight and began scorching the quarter deck. Wallace tried to lower himself by rope into the gig but was too weak to hold his own bodyweight so that the rope burned all the skin off his palms as he tumbled into the boat. With raw hands, he joined his fellow sailors in frantic bailing.

Having tied their two boats to the stern of the *Helen* by a long rope, the refugees sat, swaying with the swell, to watch the death throes of their ship and hoping also to attract a passing vessel. The fire chomped steadily through the sails, shrouds, spars and masts, until the unbalanced hulk began to roll idiotically. Helpless, Wallace had to witness his menagerie of parrots and monkeys huddle on the tip of the bowsprit to escape the heat, then turn, one after the other, to dash into the flames. The sailors rowed as close as they dared, but could rescue only one bedraggled parrot that dropped into the water from a charred bowsprit rope. As night fell, the captain ordered the boats to move further away to avoid an armada of burning timbers floating on the water.[3]

After nine days in the boats, burned and blistered and with their water almost gone, the mariners were rescued by an old

A female orang-utan. (AR Wallace, *Malay Archipelago*: Papuaweb)

cranky brig, the *Jordeson*, bound for England and weighed down by a cargo of Cuban timber. Further adventures followed when this 'rotten old tub' also nearly sank in a succession of gales. On several occasions Wallace wished himself back in the lifeboats. Eventually, on 1 October 1852, they docked at Deal after a journey of 80 days, with 4 feet of water sloshing inside the *Jordeson*'s hold. Not surprisingly, Wallace swore never to take a major ocean voyage again.[4]

Kindly headhunters

Still, Wallace was also a realist. After living for a time in London on the small insurance from his lost collection, he decided he had no choice but to voyage again to somewhere relatively unexplored in the hope of renewing both his income and scientific credibility. Having considered the Andes, the Philippines and East Africa, he decided to support himself by

collecting duplicate specimens gathered in the Malay Archipelago to sell to a number of European buyers.[5] He chose this destination, he later explained, because:

> it teems with natural productions which are elsewhere unknown. The richest of fruits and the most precious of spices are here indigenous. It produces the giant flowers of the *Rafflesia*, the great, green-winged *Ornithoptera* [butterfly], the man-like Orang-Utan, and the gorgeous Birds of Paradise ... To the ordinary Englishman this is perhaps the least known part of the globe.[6]

Even now, shipping crises continued to haunt him. His intended berth was on an Admiralty frigate, but this came to nothing because, after the outbreak of the Crimean War with Russia early in 1854, it was reassigned as a troop carrier. He then obtained alternative passage on the Peninsular and Orient steamer *Bengal*, but this entailed first making a gruelling overland trek through Egypt from Alexandria to Suez, along a desert road littered with camel skeletons. Eventually he landed in Singapore on 20 April 1854 – the launching point for an eight-year adventure which he was later to call 'the central and controlling incident of my life'.[7]

After spending a few months collecting insects in Singapore and Malacca from July to September 1854, he decided to focus his attentions for the next 16 months on the island of Borneo and its independent sultanate of Sarawak.[8] Here Sir James Brooke, the legendary adventurer and 'White Rajah', extended Wallace his patronage. Brooke had been made ruler of the eastern Borneo district of Sarawak by the Sultan of Brunei in 1841 and he was also a considerable expert on the

Sir James Brooke, the 'White-Raja of Sarawak'. (From Spenser St John,
The Life of Sir James Brooke)

orang-utan, or *mias*. With his blessing, Wallace spent much of his time in Borneo and Sarawak killing and dissecting *mias* and observing their behaviour in the wild. While living in a hut at the head of the Sarawak River in 1855 he also wrote his first portentous paper on evolution, which argued that 'every species had come into existence coincident both in time and space with a pre-existing allied species'.[9] It was to prove one of the goads that set Charles Darwin on the path to writing the *Origin of Species*.

None of this would have been possible, however, without the support of the 'primitive', head-hunting, Dyak peoples of Sarawak, whom Wallace found even more 'agreeable' than the wild Huapes Indians of the Amazon.[10] They carried him to remote headwaters in their canoes, collected specimens for him, including the fearsome orang-utan, and they offered him generous hospitality. He slept unarmed in their communal houses – with rows of smoked heads hanging from the roof – and felt himself safer from crime than anywhere in Europe. 'The more I see of uncivilized people, the better I think of human nature on the whole' he declared.[11]

Wallace's Line

Soon after this, Wallace moved from partial to complete dependence on the local peoples of the archipelago. He had intended to visit Macassar, the capital of the island of Celebes (Sulawesi), one of the least-explored of areas, but with his usual jinx he missed his ship by a day. The monsoon was now

Orang-utan attacked by Dyaks. (AR Wallace, *Malay Archipelago*: Papuaweb)

blowing in the teeth of his intended course, so he was urged to visit Bali or Lombok, and from there hope to catch a passage to the Celebes. Luck and ships were not a combination Wallace trusted, and he had no particular interest in visiting Bali because it was extensively cultivated, but he had little choice. He reached Bali on 13 June 1856, after a 20-day passage in a schooner that exemplified the multicultural East: the *Kembang Djepoon* was owned by a Chinese merchant, captained by an Englishman and manned by Javanese sailors. After two days on Bali, he decided to cross a 24-kilometre strait to the island of Lombok in search of a ship, but it had just left. Wallace's boat curse had struck again.

Though it took Wallace a few weeks to realise it, the gods of chance had actually taken pity on him. Finding few birds and insects around the town of Ampanam in Lombok, and accompanied by his young Malay assistant Ali and a Malaccan shooter, he caught a native out-rigger for a day's rowing to the southern extremity of the bay, where the wilder country was said to host flocks of birds. Flat valleys and open plains rose up to steep volcanic hills 'covered with a dense scrubby bush of bamboos and prickly trees and shrubs'. But it was the bird life that proved a shock: 'I now saw for the first time the many Australian forms that are quite absent from the islands westwards'.[12] With these understated words, Wallace announced one of the most important scientific discoveries of his life.

Australian bird species were everywhere. The white cockatoo, with its arrogant sulphur yellow crest, was impossible to miss.[13] Less flamboyant but much stranger was the *Megapodius gouldiae*, a brown, hen-like bird with long orange feet

Ali, Wallace's Malay assistant. (AR Wallace, *My Life*)

and curved claws. Rather than hatching its eggs with body-heat, it scratched up a huge mound, using sand, soil, shrubbery and any rubbish it could find. Inside this, it buried a clutch of brick-red eggs to hatch on their own. Wallace was especially intrigued because this scrub fowl belonged to a family of birds found nowhere else but Australia and its surrounding islands. What on earth was it doing here in the heartland of South-East Asia, yet absent from Bali only 24 kilometres away? The same held for many other birds: large green pigeons, kingfishers related to the 'great Laughing Jackass of Australia' (kookaburras), iridescent green bee-eaters, shy multi-coloured ground thrushes, little crimson and black flower-peckers, metallic-coloured king crows, golden orioles and 'fine jungle cocks'.[14]

Wallace quickly realised that he had stumbled on a great natural boundary – the geographical line that marked the division between the zoological regions of India and Australia. In zoological terms, India (Asia) and Australia faced each other across a 24-kilometre moat.[15] While Asia contained numerous large mammals, Australia had 'scarcely anything but marsupials', none of which were found in Asia. While Australia was the richest continent for parrots, Asia was the poorest; and so on.

A later letter from Wallace to his naturalist friend Henry Bates spelt out the implications bluntly:

In this Archipelago there are two distinct faunas rigidly circumscribed, which differ as much as those of South America and Africa, and more than those of Europe and North America:

The Papuan lorikeet, one of the Australian bird species of Lombok – the origin of Wallace's Line. (John Gould, hand-coloured lithograph, in *Birds of New Guinea and the Adjacent Papuan Islands*, 1875–1888: National Library of Australia)

CHARMOSYNA PAPUENSIS,

W.Hart. del. et lith.

Walter imp.

yet there is nothing on the map or on the face of the islands to mark their limits. The boundary line often passes between others closer than others in the same group. I believe the western part to be a separated portion of continental Asia, the eastern the fragmentary prolongation of a former Pacific continent.[16]

Six years later Thomas Huxley would give this boundary, as it extended up from Lombok past the Philippines and Timor, the name 'Wallace's Line'. A modern authority calls it 'the boldest single mark ever inscribed on the biogeographical map of the world'.[17]

Life on a *prau*

From Lombok, Wallace eventually caught a schooner to Macassar, but decided to leave in December 1856 because the wet season had set in. Black clouds invaded the sky, and driving rain was turning the fields into duck ponds. It became imperative to find a drier region for collecting. After considering the myriad possibilities reachable from this great native trade emporium, Wallace opted for the Aru Islands off the south-west coast of New Guinea. Here, the weather would remain suitable for months yet.

Wallace had long regarded this area as the '"Ultima Thule" of the East' because it stretched beyond the reach of European control.[18] From Aru, indigenous traders provided the connoisseurs of Europe and the East with the extravagant luxuries of tortoise-shell, mother of pearl, edible birds nests, dried trepang (sea slug) and the feathers of the fabled bird of paradise. These last were a particular prize. Although bird of paradise feathers had been reaching Europe for centuries, naturalists complained that the skins and carcasses were invariably mangled. Often their legs had been chopped off for ease of transport, spawning a myth that the birds remained in perpetual flight. As a result, Carl Linnaeus named the best-known species *Paradisea apoda*. By 1856, the birds were at least acknowledged to have legs, but no European had seen or studied them in the wild – a challenge that Wallace found irresistible.

For a man with Wallace's boating record, going to Aru was a bold decision. He would have to take a 1600-kilometre voyage in a Malay *prau*, and to live for six or seven months among 'lawless traders and ferocious savages'. He would need to leave Macassar with the monsoon in December or January, and to return when the winds switched direction some six months later. Making the Aru trip was regarded 'as a rather wild and romantic expedition' – even by Macassar peoples.[19]

When he at last clambered on the *prau* at daybreak on 13 December 1856, Wallace found himself aboard a ship that inverted every structural and social principle of British maritime practice. For a start the owner-captain, a mild-mannered Javanese half-caste, asked Wallace to decide his own fee when the return voyage was completed. This same captain, Wallace observed, never shouted or flogged his sailors, wore only trousers and a headscarf, and dined with the Bugis (creole Chinese) members of his crew – even though

A Malay *prau*. (Lithograph from François Edmond Pâris,
Essai sur la Construction Navale de Peuples Extra-Europeans, 1843:
Australian National Maritime Museum)

Entomological specimens collected by Wallace. (Natural History Museum, London)

they were technically criminals working off their debts by a stint of sailing. Much of the time, too, the captain ignored his compass, being content to maintain a true course by watching the swell of the sea.

The first mate, an old man known as the *jurugan*, seemed equally unlikely: he spent most of his time chanting 'Allah il Allah', and beating time with a small gong.[20] Members of the piratical-looking crew felt free to give their opinions every time the boat tacked, manoeuvres that were therefore accompanied by a babble of 'orders ... shrieking and confusion'.[21] At any given time, only a quarter of the crew were working; the rest dozed in the shadow of the sails, chatted and chewed betel in small huddles, or engaged in domestic chores like carving knife-handles and stitching shirts.

As for the ship, a giant junk-like sailing canoe that drew around 70 tons, Wallace at first thought it 'an outlandish craft'. It had been built by Ke Islanders out of planks using no nails, and it carried sails made of matting. The bow, which should have been high, was the lowest point of the boat; the rudders were situated amidships on cross-beams rather than at the stern, and they were held in place only by slings of rattan and the friction of the sea. In an exact reverse of British rigs, the longer side of the mainsail was mounted high in the air and the short side hauled down onto the deck. The tillers entered the boat through two square openings at the rear, which Wallace discovered to his alarm were only three foot from the surface of the water and completely open to the hold, 'so that half a dozen seas rolling in a stormy night would completely swamp us'. The *prau*'s 'wilderness' of rattan

and bamboo rigging, yards and spars seemed in a permanent tangle. The clock for measuring time turned out to be a half-coconut shell floating in a bucket, with a small hole in the shell designed to seep in water at a calculated rate. And Wallace's own cabin was a small 4-foot-high thatched hut on the deck, with a reed entrance and split bamboo floors.[22]

Yet it worked. Wallace had never been more comfortable or relaxed on a long sea voyage. The combination of bamboo, palm thatch, coir rope and vegetable fibres in his little 'snuggery' not only managed to keep him dry, but also smelt so sweet and natural that he was reminded of 'quiet scenes in the green and shady forest'. No stink of paint, tar, varnish, oil or grease assailed his nostrils.[23] Soon they were bobbing through the waves at a steady 5 knots, while he ate vegetables and freshly caught shark, and listened to a murmur of conversation and prayers from the crew. When they entered the Moluccan Sea at night, he was entranced 'to look down on our rudders from which rushed eddying streams of phosphoric light gemmed with whirling sparks of fire'. Despite comprising fifty 'wild, half-savage looking fellows' of several different tribes and tongues, the crew neither quarrelled nor fought. Wallace doubted whether Europeans would have behaved so well 'with as little restraint on their actions'. Even the coconut husk clock never varied from Wallace's chronometer by more than a minute. As he watched flying fish skim through the air on 100-metre flights, rising and falling like graceful swallows, it was as if the *prau* sailed in harmony with the sea.[24]

New Year's Eve was celebrated at their first port of call, the Ke Islands, a mass of coloured limestone rocks, jutting peaks

Ornithoptera priamus. (Natural History Museum, London)

and pinnacles, tall screw pines and Liliaceae. Below these were little bays and inlets, with beaches of 'dazzling whiteness' and 'water ... as transparent as crystal'. Wallace found it 'inexpressibly delightful' to be 'in a new world'. On seeing his first Papuan traders crowding around the *prau*'s Malay crew, he decided that these were 'two of the most distinctly marked races that the earth contains'.[25] The Papuans, 'intoxicated with joy and excitement', sang, shouted and shoved, touching everyone and everything; the Malays stood back, dignified, constrained, and rather affronted by such violations of their social protocol. During their four-day stay, Wallace collected a tally of 13 species of birds, 194 of beetles and three of land shells – his first specimens from within the Moluccas and New Guinea region. They included ruby and emerald beetles, splendid scarlet lorikeets and large, handsome butterflies.[26]

On 8 January 1857, they anchored at the small trading settlement of Dobbo on the Aru island of Wamma. It had been the most enjoyable voyage of Wallace's life. The Macassar *prau* and its crew became a symbol to him of all that was delightful about the archipelago. Reflecting afterwards, he decided that the freedom from excessive restraint, absurd dress codes and hierarchical pretensions of European ships had fuelled the pleasures of the voyage. As a result:

> the crew were all civil and good tempered, and with very little discipline everything went on smoothly ... so that ... I was much delighted with the trip, and was inclined to rate the luxuries of the semi-barbarous prau as surpassing those of the most magnificent screw-steamer, that highest product of our civilization.[27]

Local house in Wokan, Aru Islands, where Wallace lived. (AR Wallace, *My Life*)

Natives shooting at the great bird of paradise. (AR Wallace, *Malay Archipelago*: Papuaweb)

As in the Amazon, using native sailing craft underscored his conviction that a naturalist could best understand new places by sharing the everyday travelling life of their indigenous inhabitants, with all the accompanying perils, hardships and joys.

The traders of Aru

The same held for living on land. Having set himself up at Dobbo in a thatched bamboo shed on a spit of sand that merged with the beach, he turned his attention to the luxuriant forest behind. Chasing along the paths with his Malay assistant Ali in support, he discovered a collector's wonderland. Within a few hours he captured 30 new species of butterfly and within three days, the queen of them all – the great bird-winged butterfly *Ornithoptera Poseidon*:

> I trembled with excitement as I saw it coming majestically
> towards me, and could hardly believe I had really succeeded in
> my stroke till I had taken it out of the net and was gazing, lost
> in admiration, at the velvet black and brilliant green of its wings,
> seven inches across, its golden body and crimson breast.

It was like a gem shining in the gloom: 'The village of Dobbo held that evening at least one contented man'.[28]

After being prevented for some weeks from venturing into the interior by news that a fleet of pirates from Maguidaoa had been attacking and looting vessels near Dobbo, Wallace eventually procured a smaller *prau* to take him to the centre of the island. Here many locals specialised in hunting the

Dobbo in the trading season. (AR Wallace, *Malay Archipelago*: Papuaweb)

bird of paradise by inserting a cup-size conical wooden cap on the end of their arrows so as not to damage the exquisite plumage.

Even so, the birds were hard to come by. After four months of delays and false leads, Baderoon, Wallace's Malay cook and shooter, at last appeared one evening with a specimen. Most of the little bird's plumage was 'an intense cinnabar red', while the velvety feathers on its head shaded into glossy orange. From the breast downwards was a pure silky white crossed with a band of 'deep metallic green'. The same green surrounded the eye, contrasting with a vivid yellow bill and cobalt blue legs. From under the wings came 'tufts of greyish feathers terminated by a broad band of intense green'. These feathers could fan out in a double curve, while the two

The king bird of paradise. (John Gould, hand-coloured lithograph, in *Birds of New Guinea*: National Library of Australia)

CICINNURUS REGIUS,

J. Gould & W. Hart del et lith

Wallace as an old man. (Roger Remington, oil on canvas: By permission of the Linnean Society, London)

middle feathers of the tail took the form of slender wires 13 centimetres long, which webbed at the end into 'a pair of elegant glittering buttons'.

Wallace wrote that it was 'the most perfectly lovely of the many lovely productions of nature', and he was the first man ever to send an unblemished specimen to Europe. He was also the first man to describe the flamboyant courtship displays performed by the male birds by fanning out their wings and plumes to entice female partners, although many naturalists were to regard this story as no less fanciful than the legend that birds of paradise were born without legs.[29]

Wallace had to admit that, on the surface, the traders who supported him on Aru were an unlovely-looking lot. The 500 people who lived in Dobbo were undoubtedly 'types whom one is told have the worst reputation for morality – Chinese, Bugis, Ceramese, and half-caste Javanese, with a sprinkling of half-wild Papuans from Timor, Babber and the other islands'. Living as he did in a palm-leaf hut with an open entrance ought then to have been risky. Instead, he felt safer than when under the protection of London's metropolitan police:

> This motley, ignorant, bloodthirsty, thievish population live here without the shadow of a government, with no police, no courts, and no lawyers; yet they do not cut each other's throats, do not plunder each other day and night; do not fall into the anarchy such things might be supposed to lead to ... It puts strange thoughts into one's head about the mountain-load of government under which people exist in Europe.[30]

On 2 July 1857, after one of the most successful collecting stints of his life, Wallace departed the Aru Islands forever, this time sailing in an armada of 15 *praus* in order to deter pirates. He carried a collection of 9000 specimens of 1600 distinct species, and, thanks to the smoothness of the voyage, he did not lose one. After all his previous misfortune, a little self-satisfaction was pardonable:

> I had made the acquaintance of a strange and little-known race of men; I had become familiar with the traders of the East; I had revelled in the delights of exploring a new flora and fauna, one of the most remarkable and least-known in the world; and I had succeeded in the main object for which I had undertaken the journey – namely to obtain fine specimens of the magnificent Birds of Paradise and to be enabled to observe them in their native forests ... it is still the portion of my travels to which I look back with the most satisfaction.[31]

By the time Alfred Wallace eventually left the Malay Archipelago, after eight years of collecting, he had travelled on some 70 similar expeditions, mostly in Malay *praus*, covering an overall distance of 22 500 kilometres. From the man who hated sailing in ships, there could be no greater testament of esteem.

measuring and mapping new worlds

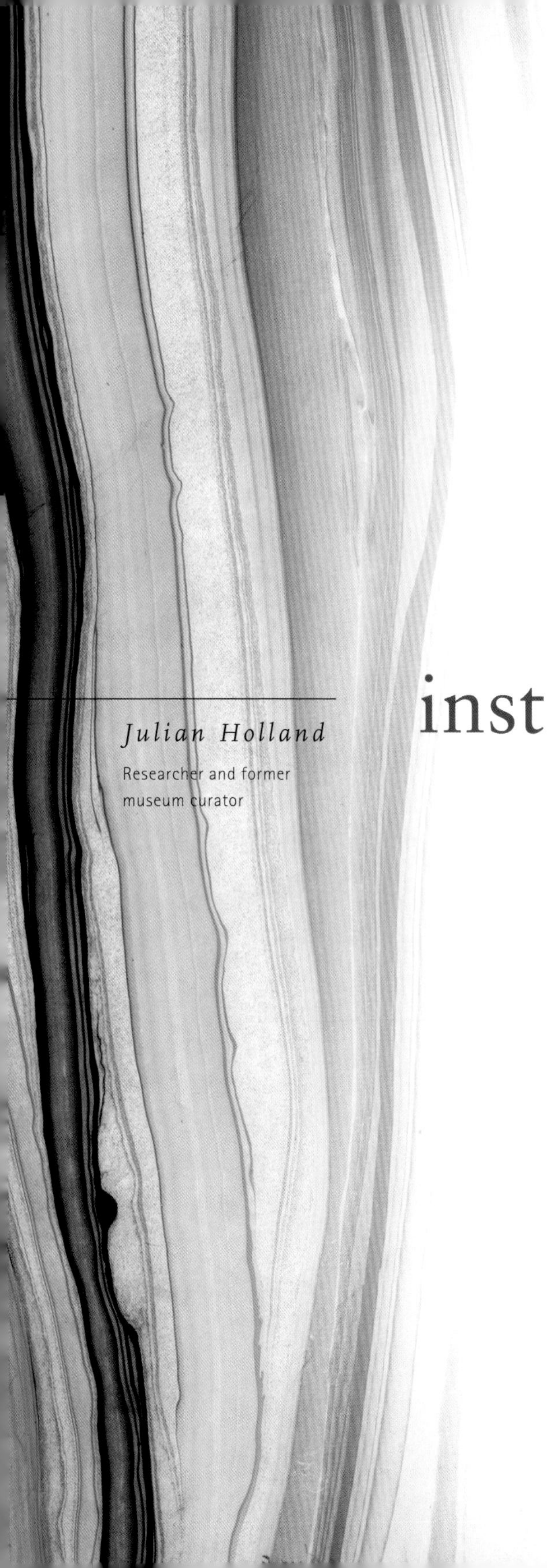

instruments and expeditions
FROM THE *BEAGLE* TO THE *CHALLENGER*

Julian Holland
Researcher and former
museum curator

Some years after their return, John Lort Stokes recalled the cramped working arrangements he had shared with Charles Darwin: 'We worked together for several years at the same table in the poop cabin of the *Beagle* during her celebrated voyage, he with his microscope and myself at the charts'.[1] The scene epitomises the instrumental requirements of the two men. For Darwin, the young gentleman-naturalist, the microscope was one part of an array of equipment that included a rifle, pistols, his geological hammer, nets, bottles and pins, as well as a shelf of reference books – Lyell's *Principals of Geology* foremost among them. For Stokes, on the other hand, behind his hydrographic chartwork lay a storehouse of scientific instruments.

A great many different types of precision instrument played a part in scientific voyages in the nineteenth century. When the *Beagle* sailed in December 1831 under the command of Robert FitzRoy, it was extremely well equipped by the standards of its day. 'Considering the limited disposable space in so very small a ship, we contrived to carry more instruments and books than one would readily suppose could be stowed away in dry and secure places', FitzRoy remarked, 'and in a part of my own cabin twenty-two chronometers were carefully placed'.[2] The *Beagle* was a tiny vessel, only 242 tons, with a total complement of 74, including Darwin among the supernumeraries. Importantly, among these was an instrument maker, George James Stebbing, whom FitzRoy had hired to attend to the chronometers and repair instruments.

Forty years later, HMS *Challenger* set out on an expedition to investigate the physical, chemical and biological character of the oceans in a voyage that was to last more than three years. The instruments she carried show how much the nineteenth century was a period of great advances in scientific knowledge and techniques. The *Challenger* was a much larger vessel than the *Beagle*, a steam-assisted corvette of 2306 tons, under Captain George Nares, a veteran of Arctic exploration and nautical surveying. In addition to 260 officers and crew there was a civilian scientific staff of five, led by Charles Wyville Thomson, professor of natural history at the University of Edinburgh. The *Challenger* was specially fitted out with zoological and chemical laboratories, and a deckhouse was erected at the stern for the messy work of the naturalists. She carried a great variety of instruments and apparatus for sounding, trawling, dredging, taking deep-sea temperatures and analysing the material gathered, as well as all the usual navigational, meteorological and geomagnetic instruments.

Instrument making

By the end of the eighteenth century, London was the unrivalled centre of scientific instrument-making in Europe. This was driven in no small part by the needs of navigation as Britain's maritime activity expanded across the globe. Changes in land use, the construction of canals and massive building programs also drove the market for land surveying instruments. These activities stimulated the invention and manufacture of improved drawing and calculating instruments. Rising wealth and literacy created a growing consumer culture, and an interest in the new scientific knowledge was encouraged by itinerant lecturers who demonstrated the new ideas with 'philosophical' instruments.

The wide range of instruments used in technical occupations and for rational entertainment created a dynamic instrument manufacturing sector and provided the economic base for developing the high precision skills used in the production of expensive made-to-order astronomical instruments. The use of those instruments at Greenwich Observatory near London, and a growing number of other observatories around the world in the early nineteenth century, established an increasingly detailed knowledge of the positions of stars that were used in navigation.

Not only were the scientific activities of nineteenth-century voyages of discovery dependent on the technical infrastructure of instrument-making developed in Britain in the eighteenth, but those scientific purposes also grew out of the expanding scientific culture of that century of which instrument making was a part.

Navigation and hydrography

The principal instruments used for navigation in the early nineteenth century were octants and sextants. These were used to measure the angle of the sun or known stars above the horizon. Such observations made it straightforward to

The zoological laboratory on the *Challenger*. Four microscopes can be seen on the central bench while around the side are numerous specimen bottles and two shelves of reference books. (CW Thomson, *Report on the Scientific Results of the Voyage of HMS* Challenger, 1885)

The artificial horizon consisted of a tray of mercury covered by a glazed hood. This enabled the angle of the sun or a star above the horizon to be measured with a sextant. (© National Maritime Museum, Greenwich, London)

development of the mechanical dividing engine by the leading London instrument maker, Jesse Ramsden.

Where the horizon of the sea could not be seen, for instance when taking observations on land, an artificial horizon was used. This consisted of a small tray in which a pool of mercury provided a reflective surface. The 'horizon' was half-way between the star and its reflection in the mercury. To protect the surface of the mercury from wind disturbance, a metal A-frame hood with two windows was placed over the tray.

The *Beagle* carried a considerable number of sextants, and also some 'full circle' instruments. For obtaining latitudes, as FitzRoy recorded:

> I was always anxious to get many results, not only by one observer, or instrument, but by several observers, and different instruments. It sometimes happened that there were six observers seated on the ground, with as many different instruments and horizons, taking the sun's circum-meridian altitudes, or observing stars at night. Where so many were working against one another, errors were soon detected, either in observation or in computing.[3]

Three of the sextants had been specially made for FitzRoy by another London maker, Nathaniel Worthington, who followed in the tradition of Ramsden's workshop. These had additional horizon glasses which enabled observations of the sun in low latitudes such as the Galapagos Islands: 'The contrivance was my own, and found to answer', FitzRoy dryly remarked.[4]

Working in severe conditions created difficulties for some

determine the latitude of a ship – how far it was north or south of the equator. Longitude – the east-west position – was more difficult, but the problem was solved by the invention of the marine chronometer.

The sextant was generally a smaller but more accurate instrument than the octant. It was made of metal, rather than the wood and bone of the octant, with a finely inscribed scale on an inlaid strip of silver (or sometimes gold or platinum). The accuracy of the reading depended in part on the precision of the scale. Knowing exactly where a ship was could mean the difference between hitting or missing a reef. The accuracy with which instrument scales could be divided was greatly improved in the late eighteenth century by the of the instruments. During the exploration up the Santa Cruz River in Patagonia in April 1834, frost affected the instruments: 'while observing the moon's meridian altitude, dew was deposited so fast upon the roof of the artificial horizon, and froze there so quickly as it fell, that I could hardly make the observation'.[5] The brass of FitzRoy's sextant also contracted so much in the cold that it affected the silvering on the index glass.

The *Challenger* also carried several sextants, a repeating circle and two artificial horizons. The *Beagle* and the *Challenger* both carried numerous other instruments – telescopes, protractors, parallel rulers, proportional compasses and so on – used in seamanship, navigation and chartmaking.

Barometers

The barometer was developed in the second half of the seventeenth century. This consisted of a glass tube containing mercury. The mercury rose or fell with changes in atmospheric pressure, and it was soon realised that these changes presaged improving or worsening weather. Barometers were increasingly available for domestic use by 1700. A glass tube filled with mercury, even when mounted on a rigid wooden backplate, was a fragile object vulnerable to damage by the motion of a ship at sea. Mercury is so heavy that even its rapid movement could shatter the glass. But by restricting the bore of the tube, the rate of movement of the mercury could be slowed, reducing the risk.

The first marine barometers were produced in the 1770s and were reasonably standardised by the time the *Beagle* was being readied for its second voyage. Even so, these were still hand-crafted products and the quality of manufacture varied. As FitzRoy noted, 'due regard should be had to the goodness of the instrument, as some barometers, used in ships, differ from others even tenths of an inch'.[6] The *Beagle* carried a number of marine barometers, including a deck barometer as well as a cabin barometer. The latter was made by the prominent London instrument-maker Thomas Jones. It was 'an excellent marine barometer' with an iron cistern, 'sent by water from the maker's hands' to Plymouth.[7] Suspended in FitzRoy's cabin, it took the principal barometrical readings made throughout the voyage. The purpose of these readings, which were taken at set hours each day, was not just to warn of coming storms but to establish the typical atmospheric pressure in different regions of the sea.

Because the marine barometer was made less responsive to atmospheric changes by the constriction of the tube, another instrument, the sympiesometer, was also carried on ships, including the *Beagle*. This was smaller than a mercury barometer, atmospheric pressure working against oil and gas in a glass tube. Developed by Alexander Adie in Edinburgh a decade or so before the *Beagle* sailed, the sympiesometer was familiar to FitzRoy:

> I must here add one word in favour of the barometer, or sympiesometer. Every material change in the weather is foretold by these invaluable instruments, if their movements

are tolerably understood by those who consult them, and if they are frequently observed.[8]

According to Darwin, there were five sympiesometers on the *Beagle*.[9]

As air pressure varies with altitude, barometers can also be used to measure height above sea level. The standard barometer had a scale range of 27 to 31 inches of mercury. Mountain barometers had extended scales to account for lower pressures. Darwin several times used a barometer for altitude measurement. For example, FitzRoy and Darwin each carried a mountain barometer on their expedition up the River Santa Cruz in April 1834. It has been suggested that Darwin purchased a barometer himself before embarking on the *Beagle*, but there is no clear evidence for this and it seems more likely that he used one from the ship's stock of instruments.[10]

The *Challenger* was also well equipped for measuring barometric pressure throughout her voyage. She carried three marine barometers and two mountain barometers of the mercury-in-glass pattern. But a metre-long glass tube containing mercury was never a convenient means of measuring altitude, especially on rough terrestrial journeys. It was therefore a great advantage when a much smaller, more robust instrument, the aneroid barometer, was developed in the 1840s. This worked by air pressure acting on a sealed metal chamber, the movement of which operated a needle on a dial. The *Challenger* carried two standard aneroid barometers and three pocket aneroids.

Thermometers

Like the barometer, the thermometer was developed in the seventeenth century. The marine barometer usually had an attached thermometer, which enabled the temperature of the barometer – then measured in Fahrenheit – to be checked and an adjustment made to the reading of the height of the mercury to account for the effects of temperature on the glass and mercury.

The *Beagle* also carried several independent thermometers, including some of the pattern known as Six's thermometer in which the mercury moved in a U-shaped glass tube, with the maximum and minimum temperatures recorded by the position of index markers moved by the mercury. Thermometers were used to record not only the air temperature but also that of the sea. The *Beagle*'s 'water thermometer', however, was lost overboard on 16 June 1834, and FitzRoy 'employed another agreeing with Six's Self Registering'.[11] When the replacement was broken in January 1835, another was substituted. It could not be assumed that the scales were equivalent on different instruments. FitzRoy noted that the substitute read nearly one degree lower than the standard.

In the early decades of the nineteenth century, there were a number of attempts to measure the temperature of the sea at various depths. This was not a regular part of the work of the *Beagle*, but some measurements were taken using self-registering thermometers. In April 1836, for example, at the Keeling Islands in the Indian Ocean, the temperature at the bottom, measured at 363 fathoms, was 'very carefully observed' to be 45°F (about 7°C).[12] In the course of this activity a self-registering thermometer was 'lost … in sounding',[13] but evidently there was another, as the following month systematic measurements were made at several depths between the surface and 420 fathoms, which showed a diminution of temperature with depth. The attrition of thermometers

Reading deep-sea thermometers on the *Challenger*. Several thermometers were set at fixed intervals along the sounding line. (CW Thomson, *Report on the Scientific Results of the Voyage of HMS* Challenger, 1885)

continued, with the water thermometer then in use getting broken in late September 1836, only days before the *Beagle* reached England again. Yet another thermometer was pressed into service.

Forty years later, the collection of 'information as to the distribution of temperature in the waters of the ocean' was one of the 'chief objects' of the voyage of the *Challenger*.[14] Besides the various thermometers used for measuring the temperature of the air and the surface water, there were numerous thermometers specially designed with an outer bulb to protect the inner bulb from the great pressure encountered at the bottom of the ocean. When taking soundings off the coast of Japan, the *Challenger* measured great depths, but this proved beyond the capacity of some of the equipment. At about 4500 fathoms, three out of four deep-sea thermometers 'sent down to these depths were crushed to pieces by the enormous pressure they had to bear'.[15] In all, 48 deep-sea thermometers were 'expended' during the course of the voyage.

As currents in the ocean meant that the temperature did not always decrease with increasing depth, the maximum and minimum thermometer did not necessarily record the actual temperature at the desired depth. In the latter part of the voyage, the *Challenger* was supplied with a new type of thermometer which could be rotated at a set depth to record the temperature at a specific depth. These thermometers were however experimental and not entirely reliable. Nonetheless, this shows how far the capacity for measuring the temperature at different depths of the ocean had advanced since the voyage of the *Beagle*.

A hydrographic surveying party at work, including an officer using a sextant and
artificial horizon (left) and (background) a row of geomagnetic instruments on tripods.
(Owen Stanley, 1840s: Mitchell Library, State Library of New South Wales)

Geomagnetism

The *Beagle* carried several compasses, including one by George Stebbing, father of FitzRoy's on-board instrument maker. Magnetic compasses had been used for navigation for centuries, but they were not altogether reliable, a problem made worse by the increasing use of iron in ships in the nineteenth century.

Compass needles, when unaffected by other influences, point to the north magnetic pole. It was long known that this was not only in a different location from geographic north, but that it moved around over time. Apart from the horizontal dimension (declination) of the earth's magnetism, instruments were also developed to measure the vertical component, known as inclination or dip, and the overall magnetic intensity. There were instruments of both kinds on the *Beagle*. The government issued a 'dip circle' by the London maker Dollond, but FitzRoy purchased another, by the Parisian Gambay, which was 'a very superior one', used for nearly all the measurements on the voyage.[16] Magnetic intensity was measured by the same Hansteen's apparatus which had been taken on the *Beagle*'s previous voyage. This was an instrument in which a magnetic cylinder was suspended. Intensity was determined by the careful measurement of time taken for 300 vibrations of the cylinder.

Geomagnetic instruments had become more sophisticated by the time the *Challenger* sailed. It carried a unifilar magnetometer, a Barrow dip circle, two Fox inclinometers and other apparatus, and during the voyage made comprehensive measurements of declination, inclination and magnetic force at over 600 separate locations.

Chronometers

The problem of longitude – the east-west position of a ship – was solved in the eighteenth century by the invention of the marine chronometer, a precision clock that could accurately maintain a reference time, say that of Greenwich Observatory, through all the motions of a ship over a long period of time. By comparing the reference time with the local time – determined by observations of the sun or known stars – it was possible to calculate the difference in meridian. The more accurately a ship could identify its position, the more readily it could avoid hazards.

By the early years of the nineteenth century, the British Admiralty was building up a stock of marine chro-

Marine chronometers were used for the determination of longitude. (© National Maritime Museum, Greenwich, London)

Captain FitzRoy gave this sextant by Worthington and Allan to John Lort
Stokes during the course of the *Beagle*'s voyage. It is possibly the 'particularly
good' sextant used to establish a chain of meridian distances around the
world. (© National Maritime Museum, Greenwich, London)

convenient for use on deck or taking ashore for comparison
at observatories. 'Few vessels will have ever left this country
with a better set of chronometers', the Admiralty Hydrographer, Francis Beaufort, remarked in November 1831.[17]

An important reason for the *Beagle* to continue across the
Pacific at the end of her South American survey, instead of
returning more directly back round Cape Horn, was to carry a
chain of meridian distances around the world. The positions
of numerous places were to be determined by careful measurements. This was done by comparing local time with that of the
chronometers. The local time 'was invariably obtained by [a]
series of equal, or corresponding altitudes of the sun; observed
by one and the same person with the same sextant, and the
same artificial horizon, placed in the same manner, both
before and after noon'.[18] This sextant was 'a particularly good
one' made for FitzRoy by Worthington and Allan. Between
observations 'it was more than usually guarded, and on no
account handled, or exposed to a change of temperature'.[19]

The difference in longitude between two places can be
stated in degrees, but also in time: 15 degrees is equivalent to
one hour. In taking observations at the same spot in Plymouth
at the beginning and end of his journey, FitzRoy should have
found that the cumulative chain of measurement added up to
24 hours. After a journey of nearly five years, however, he found
it was in excess 'by about thirty-three seconds of time'.[20] At the
end of the voyage only 11 chronometers were being relied on
for measuring meridian distances. FitzRoy could only suggest
that a larger vessel with more chronometers, making a more
direct circumnavigation, would produce a more exact result.

nometers that could be issued to navy ships. Typically a ship
would be issued with two or three chronometers, and a flagship might carry five. The *Beagle* carried 22 chronometers, 11
of them supplied by the government, a further six belonging
to FitzRoy himself, and four more provided by their makers.
The last was lent by Lord Ashburnham. The majority of these
were box chronometers, but five were pocket models, more

The *Challenger* carried nearly as many chronometers: 17 including five pocket models. Apart from refinements in temperature compensation and the use of improved lubricating oils, their basic design remained remarkably static from the late eighteenth century. Chronometers were eventually superseded by radio signals after the Great War. One chronometer purchased in 1796 was only retired from service in 1907.

Microscopes

Microscopes, by contrast, showed a dramatic technical advance in the nineteenth century, which can be at least partly illustrated in the differences in those taken on the *Beagle* and *Challenger*. In the flurry of preparations for the *Beagle*'s voyage, Darwin wrote from London to his sister Susan:

> Tell Edward to send me up in my carpet bag, my slippers, a pair of lightish walking shoes. – My Spanish books: my new microscope (about 6 inches long & 3 or 4 deep), which must have cotton stuffed inside: my geological compass. – my Father knows that: A little book, if I have got it in bedroom, Taxidermy.[21]

The dimensions Darwin gives for the microscope indicate the size of the wooden case in which it was stored. Several parts had to be assembled for use. Darwin had been exposed to the value of microscopical examination as a medical student in Edinburgh. This 'new microscope' Darwin had probably bought only a few months earlier, before the invitation to join the *Beagle*. It was a portable microscope, almost certainly the one signed 'CARY LONDON' still preserved at Down House. It is a small compound microscope: that is a microscope with an objective lens at one end of a tube and an eyepiece at the other.

Up to the early nineteenth century, the small lenses of microscopes had optical problems that were difficult to solve. The combination of lenses in compound microscopes multiplied the optical distortions. Simple microscopes – those having only a single lens – offered much less optical distortion and were more useful for scientific purposes. The Cary microscope could be used as a simple microscope, but when Darwin sought the advice of the botanist Robert Brown, it was a simple microscope of robust design by Bancks and Son that Darwin purchased. It screwed to the lid of its case to provide a stable platform, and had a large stage convenient for examining marine specimens. Several lenses provided a range of magnifications. This was the microscope Darwin used while Stokes was busy with his charts.[22]

The optical and mechanical properties of microscopes improved greatly in the 1830s and 1840s. Theoretical and practical

The single-lens microscope by Bancks & Son which Darwin purchased shortly before sailing on the *Beagle* and used throughout the voyage. (By kind permission of the Darwin Heirlooms Trust © English Heritage Photo Library)

Challenger naturalists using a Hartnack microscope (left) and a binocular microscope by Smith and Beck. (CW Thomson, *Report on the Scientific Results of the Voyage of HMS* Challenger, 1885)

advances led to the manufacture of 'achromatic' compound microscopes in the 1830s. In 1847, when he was about to embark on his long study of barnacles, Darwin bought a 'best microscope' – an achromatic compound instrument – from the London firm Smith and Beck. It was probably something similar that Huxley bought the previous year for £13 15s, prior to embarking on HMS *Rattlesnake* as assistant surgeon, though that instrument has not been traced.

Several microscopes were used by the naturalists on the *Challenger*. These illustrate how microscopes continued to develop and how techniques for preparing specimens for microscopical examination also developed as optical properties improved. In addition to five dissecting microscopes, the *Challenger* carried 11 compound microscopes, several of them belonging to the scientific staff. Four microscopes can be seen in the drawing of the zoological laboratory: a simple microscope used for dissection, and three com-

pound microscopes. The large binocular microscope was of English manufacture, but the other two were European. The institutional development of science in Germany and France in the nineteenth century created a demand for more compact instruments with high-quality optics. By the 1870s there was a growing market in England for such microscopes for serious scientific research.

The *Challenger*'s binocular microscope 'was found to be very convenient for working with low-magnifying powers, and for ascertaining the general character of the surface gatherings'.[23] But for high-magnification work, microscopes by the German-born Parisian maker Hartnack were used. Much fine detail could be seen with Hartnack's newly developed water-immersion objective lenses. Even with good light from the laboratory's window and skylights, it required some fortitude to achieve good results. The microscopist:

> by jamming his knees against the frame of the securely fixed table, could hold himself motionless. The microscopes were secured to the table at will in any position by means of small brass holdfasts. With all these arrangements for steadiness it was found possible during a gale of wind ... to work comfortably, even with very high powers.[24]

The techniques of preparing specimens for microscopical examination developed in parallel with improvements in optics. Along with the several microscopes, the *Challenger* carried numerous bottles of 'special microscopic reagents' to aid in examining specimens.

Chemistry

Both theoretical and industrial chemistry advanced rapidly in the nineteenth century. These developments were reflected in the *Challenger* being fitted out with a chemical laboratory – containing an extensive array of glassware and apparatus, including a chemical balance and spectroscopes – and a chemist being included among the civilian staff. Special collecting bottles were designed to sample ocean water from different depths, which was subsequently analysed for various properties including specific gravity. The relative density of sea water from different depths varies so slightly that it can only be measured by a very delicate instrument: *Challenger* carried four specially made glass hydrometers for this purpose.

There was one class of apparatus allied to both chemistry and optics that was not available for use on the *Beagle* – the camera. Photography was publicly announced in 1839, but it was not until the 1850s that it began to be widely practised. With the commercial availability of cameras, photographic chemicals and papers, it became practicable to take photo-

The *Challenger*'s well equipped chemical laboratory. (CW Thomson, *Report on the Scientific Results of the Voyage of HMS* Challenger, 1885)

Reading the delicate glass hydrometer on the *Challenger*. (CW Thomson, *Report on the Scientific Results of the Voyage of HMS* Challenger, 1885)

graphs on scientific expeditions. When the *Challenger* was being refitted for the voyage, a photographic workroom was constructed. While photography did not completely replace the skills of an artist, it enabled many aspects of the voyage to be recorded.

Emptying the contents of the dredge made a mess of *Challenger*'s deck but revealed the life of the bottom of the ocean. (CW Thomson, *Report on the Scientific Results of the Voyage of HMS* Challenger, 1885)

culmination of marine science in the nineteenth century, taking 20 years to publish in 50 volumes of reports. Much of this work depended on the use of microscopes to analyse the marine specimens. In the past century, new scientific techniques have been developed for both navigation and scientific research, relying on electrical measuring instruments, radio, sonar, radar, satellites and other techniques. But the scientific discoveries about the marine world made in the past century were built on the work of nineteenth-century scientists and the instruments they used.

The *Challenger* carried specially designed metal 'bottles' for collecting samples of deep-sea water. (CW Thomson, *Report on the Scientific Results of the Voyage of HMS* Challenger, 1885)

While Darwin's experiences during the voyage of the *Beagle* famously set in train the thinking that led to his theory of evolution by natural selection and changed the nature of biology, the marine surveying work in South American waters produced a legacy of charts relied on by navigators for more than a century. The *Challenger*'s scientific results were the

from sextants to submersibles

OCEAN SCIENCE TODAY

Dr Kate Wilson

Director, CSIRO Wealth from
Oceans Flagship

During the golden age of marine discovery, early European explorers surveyed Australia's coastline in their quest for the lost southern continent. Centuries later, marine scientists mapping our waters today are driven by even more compelling goals. Oceans – or more accurately, this one global ocean we share – are under pressure. Rising demands for energy, minerals and food resources, trade and shipping, coastal development and other land-uses are all threatening the biodiversity supporting these immense marine ecosystems. Global warming, ocean acidification and pollution are just some of the threats that make the work of ocean researchers more urgent than ever before.

These issues are highlighting the chasms in our knowledge of this world within our world. The collection of accurate data continues to be critically important because our oceans are so under-explored. More than 170 years after Charles Darwin recorded his discoveries during the epic second voyage of the *Beagle*, marine scientists are still discovering new specimens. In fact, between 30 and 50 per cent of the species collected on CSIRO's 'Voyages of Discovery' program are new to science, and many of those that are known are new to Australia.

The impetus to better understand and continually improve our observation and modelling of ocean conditions is yielding some of the most exciting developments in ocean science, especially through the growth of automated observation technologies which can access areas previously beyond our reach.

Deep-ocean corals like these held by CSIRO's Dr Ron Thresher provide a litmus test for scientists of changes in ocean chemistry. (Photo Bruce Miller: CSIRO)

For example, a research program using a robotic submersible to probe Tasmania's ocean depths in January 2008 demonstrated the capacities of these technologies to go where no scientist could. The sub filmed and collected fossilised corals from extinct reefs that were hundreds of thousands of years old. At depths of up to three kilometres – deeper than any previous scientific foray into Australian waters – the submersible launched from the Marine National Facility Research Vessel *Southern Surveyor* navigated the eerie cliffs and canyons of the Tasman Fracture Zone Marine Reserve. The corals' composition, to be analysed in the laboratory, will have recorded changes in sea temperature, currents and salinity over tens to hundreds of thousands of years and are expected to reveal when the reef was alive and whether its extinction was linked to climate change. This ancient history may help forecast our near future, given the dire scenarios currently unfolding which suggest that up to a third of the world's reef-building coral species are threatened with extinction due to climate change and ocean acidification.

Quest for knowledge drives modern explorers

It is not just the ability to access new areas such as ocean depths that is driving the push for automation in studying our marine environment. Of even greater importance is the ability to obtain continuous measurements – perhaps on an hourly basis – from instruments distributed throughout our oceans.

These instruments may float with the currents, or be affixed to the seabed, deployed from research vessels or 'ships of opportunity' – commercial vessels that volunteer to assist in observation programs – or even attached to marine animals.

Whatever the method of deployment, these new measurement techniques are just beginning to give us data with the appropriate frequency and geographic coverage to enable us to understand, analyse and detect changes in our oceans over the course of one, ten or even more years. This contrasts sharply with our position as little as a decade ago, when measurements were almost entirely dependent on instruments deployed from ships, which severely limited the location, frequency and accuracy of sampling.

Modern ocean research is resource intensive, and the *Southern Surveyor* voyage – along with many others in Australia – was only made possible through strong, collaborative relationships between scientists and their various institutions. The submersible was on loan from the Woods Hollow Oceanographic Institution in the United States. The scientific team included researchers from CSIRO's Wealth from Oceans National Research Flagship, Woods Hollow, the United States' National Science Foundation, and the marine division of the Australian Department of Environment, Water, Heritage and the Arts. The sheer scale of the research challenge is uniting ocean scientists from around the world to pool resources in a common mission to understand more about these dynamic environments.

Today's marine scientists are twenty-first century explorers contributing to another golden age of marine discovery: an era when it is routine to link seabeds and space satellites in order to map the ocean's mysteries. These maps are now three-dimensional computer models that use continuously updated data – capable of charting ocean conditions in the past, present and, potentially, the future. Every day, around the clock, these models draw on information about temperature, salinity, nutrients, currents, winds and more, from monitoring webs that span the globe.

Computer modelling and automated observation systems are evolving rapidly, improving our capacities to capture and process complex data in real time, to interface with existing technologies while opening doors to new applications, and to operate efficiently across broader territory. Most importantly, these tools are helping us identify and track ocean phenomena that occur over vast time scales, and forecast ocean conditions and their interactions with other natural systems. Oceans, much more than any terrestrial systems, are integrated, with a global circulation of currents that propel water flows right around the planet and drive the climate system. Thus, data delivered by these observing systems is informing crucial decisions that govern the management and conservation of both marine and terrestrial environments.

Ocean observation technologies have confirmed that small changes in marine conditions in our southern oceans, the so-called 'engine room of climate change', can have major implications on a global scale. And our capacity to monitor seemingly small details, such as changes in water temperature and salinity, could enable us to forecast and prepare for sea-level rises, droughts or cyclones, along with less dramatic, day-to-day land management issues.

Charting unknown waters

If you travel anywhere in Australia on land, you could expect to use maps that are accurate and reasonably up-to-date. In our territorial waters, there are areas where we still rely on surveys done by Matthew Flinders and his contemporaries in the early nineteenth century. This has safety implications for mariners, as evidenced by the grounding of vessels off the Western Australian coast in recent years. Mapping across many dimensions, encompassing ocean topography, ecology, biogeochemistry and much more, continues to be a core part of our research activities.

But even if some of the maps still in use are antiquated, our mapping technologies have become highly advanced.

In Western Australia's Ningaloo Reef lagoon, for example, remote sensing in the form of hyperspectral imaging – measuring the reflection of hundreds of wavelengths of light from surface features – was used in the largest such mapping study of a coral reef ever attempted. The Australian Institute of Marine Science co-ordinated the mapping, which was funded by BHP Billiton and assisted by HyVista Corporation. This information was then interpreted to describe the underwater topography and habitats of the Ningaloo lagoon over the whole marine park. This interpretation project is part of the $12 million Ningaloo Collaboration Cluster initiated by CSIRO's Wealth from Oceans Flagship and involving six universities and the Sustainable Tourism Co-operative Research Centre. The cluster is measuring the reef's biology and reef use with the aim of developing models that will help to bal-

A multibeam sonar three-dimensional image of the Dugong volcano, discovered north-east of Fiji during a 2008 *Southern Surveyor* research voyage. (Richard Arculus, Australian National University)

ance growth in tourism with local community interests and conservation of this environmentally fragile area.

Mapping is also producing spectacular discoveries about the earth's ancient topography. In previously uncharted waters north-east of Fiji in 2008, the *Southern Surveyor's* multi-beam sonar mapping system revealed several huge, active underwater volcanoes. Multi-beam sonar builds up images of the ocean depths by bouncing sound waves off the seabed. The voyage's chief scientist, Australian National University Professor Richard Arculus, compared the seabed terrain, created by extreme volcanic and tectonic upheavals, to the volcanic blisters seen on the surface of Venus. He believes the depositions of minerals such as copper, zinc, and lead may hold clues about undiscovered deposits in Australia.

In addition to acoustic technologies such as multi-beam sonar, towed underwater camera systems on research vessels such as *Southern Surveyor* provide fine-scale visual observations to map seabed habitats and communities, monitor fish behaviour and habitat use, and estimate fish distribution and abundance.

At the opposite extreme, satellite technology has for many years been used to map the world's oceans, allowing us to chart not just physical features but to measure variations in sea surface height and temperatures on a daily basis – a high frequency of monitoring which is invaluable for building meaningful models. Without satellites facilitating the sharing of huge volumes of data between countries, ocean science networks could not function.

Networks unite global observation systems

The networks of ocean science disperse information in ways that mimic the flow of global currents in bypassing international borders. But unlike those ocean currents, the data generated by ocean science networks can flow both ways simultaneously, via a latticework of organisations that support these observing and monitoring technologies.

As the breadth of Australia's ocean observation activities grows, co-ordination and co-operation between research organisations has become paramount. Australia's community of marine scientists has long recognised the need to take a national, strategic approach to developing an effective infrastructure for ocean observation systems. After extensive consultation they created a national framework: the Integrated Marine Observing System (IMOS). By 2008, ten institutions were involved in IMOS, which co-ordinates the deployment of a wide range of equipment and assembles data through 11 facilities around the country. Researchers access data through the electronic Marine Information Infrastructure located at the University of Tasmania, which also oversees IMOS.

The operations of IMOS have diverse, community-wide benefits: providing data to international meteorological services, shipping, resources, tourism and fishing industries, search and rescue activities, and other applications. In myriad ways, IMOS is a key contributor to research into climate change, and fulfils Australia's role in international programs of ocean observation.

'Rockets' of the deep

IMOS's observational networks are manifested in some intriguing forms. If you were sailing a long way from land and witnessed a 1.9-metre, rocket-shaped object topped with antennae suddenly bob up from the blue depths, you may be encountering an Argo float. These floats have revolutionised our ability to observe the huge global ocean and can access areas that research vessels would find difficult to reach, such as the Southern Ocean in winter. Functioning like underwater weather balloons, they provide data from beneath the surface that were never previously available and with a frequency that allows much more meaningful computer modelling. The floats are fitted with satellite antennae and drift at depths of between one and two kilometres. Every ten days each float ascends to the surface, measuring temperature and salinity as it rises. These data are transmitted to satellites and within 24 hours they are available to researchers. The float's new location each time it surfaces also provides information about ocean currents. The float then dives and starts a new cycle.

The Argo project is one of the 11 IMOS facilities, and also part of another international network – this one involving 18 countries and the European Union. Australia was one of the pioneers in the Argo project and launched the first ten Argo floats in the Indian Ocean in 1999. It has now launched more than 200. World-wide, the target of 3000 active floats has been achieved, each providing year-round, near real-time informa-tion on ocean conditions. The Argo network, combined with the internet, has narrowed the constraints of time and distance to allow the public to see where Argo floats have been drifting across the world's oceans and track ocean changes.

This project has produced a dramatic increase in the quantity, timeliness, spatial coverage and quality of observations. When Argo was launched in 1999, expendable bathy thermographs – ocean probes that measure subsurface ocean temperature, and which could only be deployed from ships – were providing about 2500 profiles per month across a small and globally scattered number of sites. By mid-2008, Argo floats were providing over 9000 more accurate profiles per month across a vastly greater area.

The data from Argo floats are being used to track warming of the oceans and provide a much more accurate view of how much this warming contributes to sea-level rise. Meteorological organisations all around the world rely on Argo data for their analyses, and the data are also being used in research on the El Nino Southern Oscillation effect – a natural phenomenon that links sea surface temperature in the tropics north of Australia with rainfall and drought throughout the continent.

Anatomy of a network

IMOS exemplifies the way ocean science networks function, with each organisation, from international levels down to field stations, making valuable contributions to ocean knowledge. IMOS is part of the National Collaborative Research Infrastructure Strategy. It operates through five nodes around

Australia and contributes to and is supported by at least nine international programs, including the Global Ocean Observing System.

The newest IMOS node (as of 2008) is the Great Barrier Reef Ocean Observing System, which is managed by the Australian Institute of Marine Science on behalf of a consortium of research institutes, including the Tropical Marine Network. The Tropical Marine Network's own consortium is made up of island research stations, including Queensland's Heron Island – an important site for the long-term collection of baseline oceanographic data on the southern Great Barrier Reef.

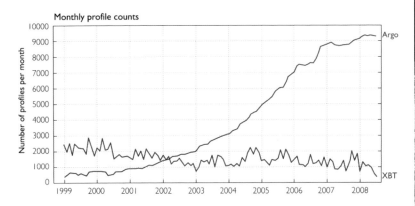

Monthly profile counts

Number of profiles provided by expendable bathy thermographs (XBTs) and Argo floats since 1999, when the Argo project was launched. (JCOMMOPS; CSIRO)

A University of Washington Argo float ready for deployment by Australian scientists aboard *Aurora Australis* in East Antarctica in 2007. (Photo Guy Williams, Antarctic Climate & Ecosystem Co-operative Research Centre)

Mobile phone network delivers data

The launch of the Great Barrier Reef Ocean Observing System took ocean observation systems to a new level by incorporating the world's first large-scale, reef-based internet protocol network. This means that high-resolution data gathered via a 'digital skin' of sensors can be transmitted using Telstra's 3G mobile phone network. The network monitors conditions over the eastern Coral Sea and Great Barrier Reef through multiple sensors deployed along the reef from Cooktown to Gladstone, measuring water temperature, salinity and nutrients. This is providing extremely detailed data that will increase the scope of its possible uses, including the development of virtual reality representations of life on the reef.

As well as the 3G network, the techno-

IMOS is using the latest advances in technology, including satellites and instruments, to observe physical and biological properties of the ocean around Australia. (Louise Bell: IMOS)

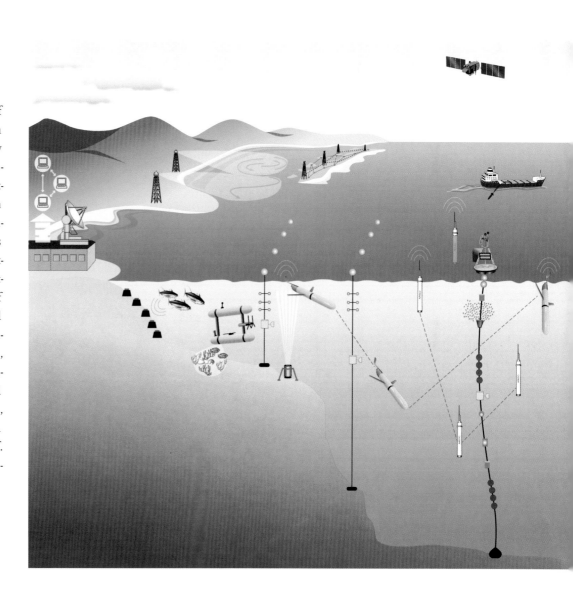

logies transmitting the data range from high-frequency coastal radar to experimental over-the-horizon microwave technology developed by James Cook University. The new microwave technology boosts our observing capacities by exploiting a physical phenomenon known as the surface humidity duct that is found above tropical waters. This 10-metre corridor sits above the water and eliminates the need for microwaves to be sent only to line-of-sight receivers, thus speeding up data gathering from remote locations on the reef.

The Great Barrier Reef Ocean Observing System's mission includes improving observations of water circulation in the Coral Sea and along the Great Barrier Reef, where the main north and south currents running along the east coast of Australia and up to Papua New Guinea are formed.

Forecasting ocean 'weather'

The significant advances in data sourced through Argo and ocean observation satellites, backed by powerful, purpose-designed computers, have made another technological breakthrough possible. The BLUElink project is an ocean modelling system that provides the first six-day forecasts of ocean temperature, salinity and currents with a focus on the southern hemisphere. Satellites and Argo floats provide BLUElink with data about surface winds, temperature, sea level, ocean currents, and sub-surface temperature and salinity. Created by CSIRO's Wealth from Oceans Flagship, the Bureau of Meteo-

rology and the Royal Australian Navy and launched in 2007, BLUElink was tested by replicating past conditions in Australia's oceans in a process called re-analysis or 'hindcasting'. For six months, scientists assimilated 15 years of quality-controlled ocean and atmospheric data, recorded since 1992, into a high-resolution ocean circulation model – the first Australian simulation of ocean conditions.

BLUElink's hindcasting capacity was one of several tools used in the exciting discovery of the wreck of HMAS *Sydney* 250 kilometres off the Western Australian coast in 2008. BLUElink re-analysis generated an ensemble of projections for November 1941, taking into account estimates of historical ocean eddies, drift and wind patterns to identify the most promising search area – modelling that proved to be right on target.

The major forecasting component of BLUElink is a global ocean modelling and assimilation system, called Ocean-MAPS. BLUElink also provided a high-resolution, three-dimensional ocean climatological atlas of average monthly temperature and salinity levels. The public is using BLUElink via the Bureau of Meteorology's website, where they can select a coastline area and watch animated seven-day forecasts predicting ocean 'weather'.

BLUElink's other applications include fisheries management, search and rescue activities, environmental protection through the modelling of threats such as an oil spill and derelict fishing nets, fuel-efficient routing and berthing for shipping, and offshore engineering and design for sectors such as oil and gas.

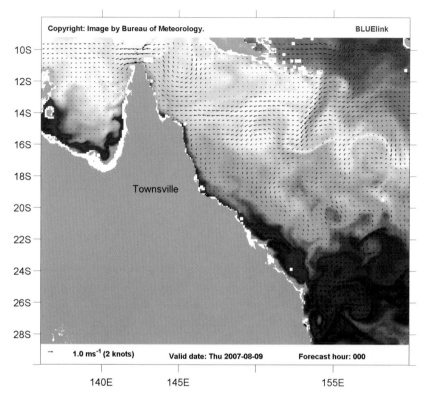

BLUElink forecast of sea surface temperatures off north-eastern Australia.
(Bureau of Meteorology)

The Royal Australian Navy is a key user of another BLUElink product, called the Relocatable Ocean-Atmosphere Model. This enables high-resolution, regional modelling that can, for example, produce a six-day forecast in areas as large as Bass Strait or the Timor Sea. It supports the navy's amphibious, submarine and mine warfare operations by, for instance, providing information on ocean conditions that substantially affect sonar behaviour.

High-tech tagging programs

Back to that yacht you were lazing about on when you bumped into an Argo float. Unbeknownst to you, another link in the data chain – a shark with a high-tech tag on its dorsal fin – was circling deep below. It was collecting data as part of a large-scale experiment aimed at balancing the maintenance of fisheries and marine ecosystems. In March 2008, deep sea sharks were fitted with acoustic tags and tracked as part of a CSIRO Wealth from Oceans Flagship project to test the conservation value of areas that were closed to commercial fishing. Fifty gulper sharks, swellsharks and green eye dogfish were tagged near Port Lincoln in South Australia. The closure to commercial fishing is intended to protect the gulper shark – a species which is severely depleted over much of its range and is nominated for protection.

These sharks may also be tracked by 24 acoustic listening stations that are part of IMOS's Australian Acoustic Tagging and Monitoring System, yet another network of about 1000 acoustic receivers that monitor tagged marine animals around Australia. The receivers can be left on the sea floor for up to seven years and upload data as often as needed. This acoustic tagging system is leading the southern hemisphere section of the Ocean Tracking Network that tracks thousands of animals around the world while building a record of climate change.

The melting sea ice associated with global warming has made many tagged animals the 'canaries in the coal mine' of ocean ecosystems. This is apparent through research into southern elephant seals, which have also been recruited as part of our ocean observing systems. Tagging of these creatures, which occupy the top of the ocean food chain, is providing evidence that the decline in sea ice may have contributed to the decline in particular seal populations.

In early 2003, miniature oceanographic sensors were glued to the fur of 85 elephant seals to track their feeding trips during the long Antarctic winter. The tags were retrieved when the animals returned to the same beach to moult, up to 10 months later. Between their tagging and their return, these seals' ocean odysseys covered thousands of kilometres and took them to depths of up to 1500 metres in their search for food.

Their tags yielded valuable data, recording each seal's position, monitoring its diving cycle with a pressure sensor, and measuring water temperature and salinity – data that is uploaded to satellite when they surface. By monitoring changes in the rate at which seals drift up or down during passive drift dives, the scientists determined where the seals were gaining fat (and becoming more buoyant) and where food was harder to find and the animals leaner.

The project provided data on a part of the ocean that is virtually unexplored, lying as it does beneath the winter sea ice where it is extremely difficult to sample from conventional platforms, such as satellites, profiling floats and ships.

The longest track was 326 days, and up to 30 000 profiles of temperature and salinity were obtained. By simultaneously recording movements, dive behaviour and oceanographic conditions, the sensors allowed researchers to examine in detail how elephant seals respond to changes in ocean conditions, including how they might be affected by climate change. This program was another high level international collaboration, this time involving scientists from Australia, France, the United States and the United Kingdom.

Fine-tuning our focus on coastlines

So far this discussion has mostly examined the importance of observation and modelling in deep ocean waters. These activities are arguably equally important, and even more challenging, as we move into Australia's coastal zones. Coastal monitoring, where the seabed has a greater influence on water movement, sometimes requires observational tools that are not at the mercy of the currents. For example, the Slocum ocean 'glider' now in use by IMOS has an inbuilt Global Positioning System and fins that control its direction. This battery-powered submersible adjusts its buoyancy to sink to depths from 200 up to 2000 metres before gliding back along pre-programmed trajectories collecting oceanographic data as it goes. It measures boundary currents like the East Australian (Southern Ocean and eastern Australia), Flinders (southern Australia) and Leeuwin (western Australia), providing information about their links to coastal ecosystems. Capable of voyages of up to 30 days, it communicates via satellite once it surfaces, and can be monitored and programmed from onshore.

The closer we get to our coastlines, the greater the complex-

An acoustic tag attached to a gulper shark as part of a CSIRO Wealth from Oceans Flagship project to test the conservation value of areas closed to commercial fishing. (CSIRO)

A tagged southern elephant seal on the Atlantic island of South Georgia. (Photo Mike Fedak, St Andrews University)

breed complacency about our damaging impact on the water environments we love. If the damage is obvious, we may feel concerned, but if it is out of sight – beyond the wave line – its significance may not arouse public attention.

Observation and modelling of coastal conditions are essential for balancing marine conservation and our many demands. But our efforts are like trying to reverse the tide when we have not yet established the baselines against which these changes can be measured. Seabed mining, offshore oil and gas drilling, commercial fishing, climate change, ocean acidification – we cannot know what changes lie ahead if we do not have baseline material and continuous monitoring to identify and track changes over oceanic timeframes.

Predictive modelling

To establish these baseline data and continue monitoring is admittedly a huge undertaking, but it is one that could gain greater momentum if we structured in a higher level of accountability for our coastlines. One path to achieving this could be a comprehensive, regular audit in the form of a 'state of our coastlines' marine environment report, to be modelled on the existing National State of the Environment Report. It would be a resource for taking a more rigorous and informed approach to making coastal policy and planning decisions, and promote greater political and community awareness about these issues. A 'state of our coastlines' report would distil all of the information from our observation and monitoring resources, including the latest advances.

ity in observing and modelling marine conditions – and the greater the expectations of those who use our research. In deep water research, these users include researchers, navies, meteorologists, and specific industries. The numbers and nature of such groups expand enormously as we focus our work closer to coastlines.

This is because Australia's population hugs the shore: most of us live within a day's drive of the coastline. Many of our industries are also based along the coast – obviously marine ones, but also numerous others. This familiarity tends to

One of these advances is BLUElink 2. Using a 'nesting' system, BLUElink 2 has a resolution of down to 2 kilometres, and it can be moved anywhere in the oceans around Australia. This allows us to zero in on nominated longitudes and latitudes, forecasting ocean conditions that take into account not just physics but also, in the future, ocean biogeochemistry. This capacity to model the interactions between the biology, chemistry and geology of a marine system would enable us, for example, to model the circulation of nutrients in these highly complex coastal zones. The BLUElink project is operating at the forefront of international modelling for ocean forecasting; our goal is to explore its potential for forecasting conditions in marine ecosystems.

Ultimately, the BLUElink project aims to evolve in ways that facilitate other modelling systems that could be applied nationally. For example, if we could develop a predictive modelling capacity that would use data on ocean currents and colour, and develop nutrient sensors that would be set in offshore coastal regions, then we would have a powerful tool to tell us what is happening in these regions in connection with human activity. It would help us to preserve the environment while making coastal planning decisions much more rigorous and uniform. National baseline modelling and ongoing monitoring would support more coherent predictive modelling and decision-making. This would help resolve controversial issues arising in projects involving, for example, dredging or effluent dispersion.

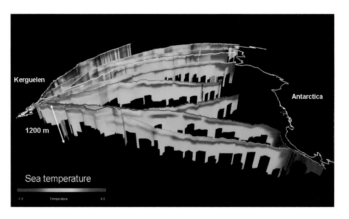

Direction and dive patterns of southern elephant seals tagged with miniature oceanographic sensors. (Martin Biuw, St Andrews University)

A big country gets bigger

Australia 'grew' by 2.5 million square kilometres in April 2008 when the United Nations confirmed Australia's entitlement to jurisdiction over an extended area of our continental shelf. Australia is probably the first country in the world to be granted such an extension. Even before it was won, we administered one of the world's largest marine territories. Now with a total of 13.6 million square kilometres to cover, are we capable of fulfilling this immense responsibility?

We have come a long way from the days when sampling our ocean occurred only intermittently, from ships, across a confined area, often at great expense but not high accuracy.

Australia is now recognised as a world leader in many areas of ocean research, despite the challenge presented by a vast coastline and small population with limited resources. For instance, the *Southern Surveyor* is our only dedicated national research vessel capable of entering deep water; Canada has significantly less territorial water to cover than Australia and has 17 blue water research vessels; Belgium, with only 40 kilometres of coastline, has three.

Quite apart from our international standing, our nation's ocean research efforts are critical to protecting Australia's national interest. The vast additions to our territorial waters have raised hopes of discovering mineral or petroleum resources – and the prospect that Australia might be making momentous decisions about our continental shelf before we know what is out there.

In the decades ahead, our nation will face many major issues that demand wise decisions – about mineral and petroleum resources, fisheries, tourism, climate change, pollution and biodiversity, to nominate a few. Only through ocean observation systems will Australia have the capacity to monitor, forecast, protect and manage its vast marine environment. Making the right choices will require a global approach that recognises that the world's many borders share just one ocean.

Australia's continental shelf jurisdiction. The area beyond 200 nautical miles, as confirmed by the United Nations, is shown in purple. (Geoscience Australia)

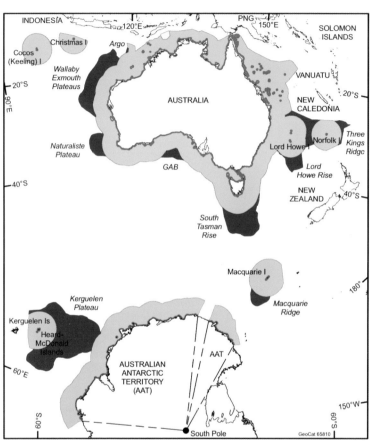

AUSTRALIA'S CONTINENTAL SHELF CONFIRMED BY THE COMMISSION ON THE LIMITS OF THE CONTINENTAL SHELF

Territorial sea and internal waters

Areas of marine jurisdiction within 200 M of Australia and its external territories

Area of Australia's continental shelf beyond 200 M as confirmed by the Commission on the Limits of the Continental Shelf

Joint Petroleum Development Area under Timor Sea Treaty 2002

Note: The areas of continental shelf depicted to the north-west of Australia reflect the terms of the 1997 maritime boundary treaty with Indonesia which has not yet entered into force.

1 nautical mile (M) = 1852m

notes

Introduction

1 Thomas Campbell, 'Ye mariners of England', in A Quiller-Couch (ed) (1907), *Oxford Book of English Verse*, Clarendon Press, Oxford, 672.

2 J Morris (1979), *Heaven's Command: An Imperial Progress*, Penguin Books, Middlesex, 22.

3 F Fleming (2001), *Barrow's Boys*, Granta, London, 1–12; C Lloyd (1970), *Mr Barrow of the Admiralty: A Life of Sir John Barrow*, Collins, London, 89–111.

4 J & M Gribbin (2003), *Fitz-Roy: The Remarkable Story of Darwin's Captain and the Invention of the Weather Forecast*, Headline, London, 80–82.

5 Quiller-Couch, *Oxford Book of English Verse*, 598.

6 C Darwin (1985), *On the Origin of Species*, ed J W Burrow, Penguin Books, London, 53.

7 B Trigger (1997), *A History of Archaeological Thought*, Cambridge University Press, Cambridge, 92.

The evolution of a tradition

1 Charles Darwin to Robert Darwin, 31 August 1831: http://www.darwinproject.ac.uk/ darwinletters/calendar/ entry-109.html.

2 Charles Darwin to Robert Darwin, 31 August 1831.

3 JC Beaglehole (ed) (1963), *Endeavour Journal of Joseph Banks 1768–1771*, Halstead Press, Sydney, 126.

4 The *Investigator* continued in service until 1810 and was then sold: see D Lyon, (1993) *The Sailing Navy List*, Conway Maritime Press, 250.

5 F Darwin (ed) (1887), *The Life and Letters of Charles Darwin Including an Autobiographical Chapter*, John Murray, London, 1: 74. The Linnean Society meeting where Darwin's and Wallaces' papers on evolution were read on 1 July 1858 was called especially to elect a new member to the society's council to fill the vacancy left by the recent death of Robert Brown: see JWT Moody (1971),'The reading of the Darwin and Wallace papers: an historical "non-event"', *Journal of the Society for the Bibliography of Natural History*, 5: 474–76.

6 GS Ritchie (1995), *The Admiralty Chart*, Pentland Press, Durham, 116.

7 His son, also Philip Gidley King, sailed as a midshipman aboard the *Beagle* under FitzRoy, and his niece, Anna Macarthur, later married Commander John Wickham of the *Beagle* in Australia.

8 David Lyon (*Sailing Navy List*, 144–46) lists 101 vessels built to the class, of which 10 foundered and 16 were wrecked. Five were fitted as survey vessels: *Chanticleer, Beagle, Fairy, Saracen* and *Scorpion*.

9 RD Keynes (ed) (2001), *Charles Darwin's* Beagle *Diary*, Cambridge University Press, Cambridge, 413.

10 CR Darwin (1842), *The Structure and Formation of Coral Reefs*, Smith Elder and Co, London.

11 JL Stokes (1846), *Discoveries in Australia*, T and W Boone, London, 6.

12 These were Lieutenant Stokes, Surgeon Bynoe, Boatswain Thomas Sorrell and two marines. Although the *Beagle* did not carry a naturalist on this voyage, Bynoe and the other officers sent natural history collections to England where they were described by Sir John Richardson, John Gould and Adam White. After Wickham was invalided out of the service in 1841, Stokes assumed command of the *Beagle*.

13 Included amongst these was another ten-gun brig, HMS *Britomart*, commanded by Owen Stanley and sent out to assist the settlement of Port Essington in northern Australia. Stanley was a midshipman on the South American station (HMS *Ganges*) and had served briefly aboard HMS *Adventure* under Phillip Parker King. HMS *Fly* and *Bramble* were involved in surveys of the Great Barrier Reef and southern New Guinea between 1842 and 1845. Included in their personnel were the geologist Joseph Beete Jukes and the zoologist John MacGillivray.

14 JD Hooker (1855–59), *Flora Tasmaniae*, part 3 of *The Botany of the Antarctic Voyage of HM Discovery Ships* Erebus *and* Terror *in the Years 1839–1843*, London.

Art and exploration

1 R Joppien and B Smith (1987), *The Art of Captain Cook's Voyages*, Oxford University Press, London, vol 3, 2.

2 J Hawkesworth (ed) (1773), *Account of the Voyages Undertaken by the Order of His Present Majesty for Making Discoveries in the Southern Hemisphere*, W Strahan & T Cadell, London, plate 20.

3 W Tench (1789), *Narrative of the Expedition to Botany Bay*, J Debrett, London, 126.

4 R King (ed) (1990), *The Secret History of the Convict Colony: Alexandro Malaspinas Report on

the British Settlement of New South Wales, Allen & Unwin, Sydney, 159.
5 Mitchell Library PXA 563.
6 George Tobin (1797), *Journal of HMS* Providence, p 82: see Mitchell Library A562. The dolphin fish also impressed William Bligh, his captain: see Mitchell Library PXA 565, f 4.
7 Mitchell Library PXE 949.
8 Mitchell Library PXD 226.
9 Mitchell Library PXD 11; Mitchell Library PXD 72.
10 Mitchell Library PXA 1005; PXC 279 & PXC 280; and PXC 281.
11 Dixson Library DL Pxx 2.
12 Warwick Hirst (2004), *Upon a Painted Ocean: Sir Oswald Brierly*, State Library of New South Wales, Sydney, 2.
13 Charles Wilkes (1845), *Narrative of the United States Exploring Expedition, During the Years 1838, 1839, 1840, 1841, 1841*, Wiley & Putnam, London, xvii.
14 R FitzRoy (1839), *Narrative of the Surveying Voyages of His Majesty's Ships* Adventure *and* Beagle, *between the Years 1826 and 1836*, Henry Colbum, London, 34.
15 Dixson Library DL Px 13.

Well salted in early life
1 T Huxley (1854), 'Science at sea', *Westminster Review*, 1 January, no 5, 98–119.
2 RD Keynes (ed) (2001), *Charles Darwin's* Beagle *Diary*, Cambridge University Press, Cambridge, 16, 35; Janet Browne (1996), *Charles Darwin: Voyaging*, Pimlico, Sydney, 178.
3 L Huxley (ed) (1978), *Life and Letters of Joseph Hooker*, Arno, New York, vol 1, 70.
4 J Huxley (ed) (1937), *TH Huxley's Diary of the* Rattlesnake, Doubleday, New York, 15.
5 Huxley, 'Science at sea', 99, 108.
6 Keynes, *Charles Darwin's* Beagle diary, 380, 442.
7 L Huxley (ed) (1900), *Life and Letters of Thomas Henry Huxley*, Macmillan, London, vol 1, 44; T Huxley to H Heathorn, 14 May 1848, Thomas Huxley/Henrietta Heathorn correspondence 1847–1854, Thomas Henry Huxley papers, Imperial College of Science and Technology, contained in the microform collection, Darwin, Huxley and the Natural Sciences, unit 2, mfm 054; Huxley, *Diary of the* Rattlesnake, 110; Huxley, 'Science at sea', 105, 112–13.
8 MJ Ross (1982), *Ross in the Antarctic: The voyages of James Clark Ross in HM's Ships* Erebus *&* Terror *1839–43*, Caedmon, Whitby, 113–14.
9 Huxley, 'Science at sea', 106.
10 Huxley, 'Science at sea', 111.
11 N Barlow (1969), *The Autobiography of Charles Darwin 1809–1882*, Norton, New York, 65–66; Browne, *Charles Darwin: Voyaging*, 192; J Hackforth Jones (1980), *Augustus Earle: Travel Artist*, Scolar Press, London; S de Vries-Evans (1993), *Conrad Martens: On the* Beagle *and in Australia*, Pandanus, Brisbane; J & M Gribbin (2004), *Fitz-Roy: The Remarkable Story of Charles Darwin's Captain and the Invention of the Weather Forecast*, Headline Publishing, London, 128.
12 Huxley, *Life and Letters of Joseph Hooker*, vol 1, 57; M Allan (1967), *The Hookers of Kew 1785–1911*, Joseph, London, 116–117.
13 Quoted in MP Winsor (1976), *Starfish, Jellyfish, and the Order of Life*, Yale University Press, New Haven and London, 74.
14 Huxley, *Life and Letters of Thomas Henry Huxley*, vol 1, 33–34; Huxley, 'Science at sea'; Huxley, *Diary of the* Rattlesnake, 51.
15 Keynes, *Charles Darwin's* Beagle *Diary*, 417–18; P Armstrong (1991), *Under the Blue Vault of the Sea: A Study of Charles Darwin's Sojourn in the Cocos-Keeling Islands*, Indian Ocean Centre for Peace Studies, Nedlands, 88–105.
16 Huxley, *Life and Letters of Joseph Hooker*, vol 1, 487–88; Ross, *Ross in the Antarctic*, 210–22; Allen, *The Hookers of Kew*, 219; Keynes, *Charles Darwin's* Beagle *Diary*, 446.
17 Huxley to Hooker, 15 November 1888, *Life and Letters of Thomas Henry Huxley*, vol 2, 211.
18 Browne, *Charles Darwin: Voyaging*, 530–31.

A Gunn and two Hookers
1 AM Buchanan (1990), 'Ronald Campbell Gunn (1808–1881)', in PS Short (ed), *A History of Systematic Botany in Australia*, Australian Systematic Botany Society, Melbourne, 184.
2 JD Hooker (1860), *On the Flora of Australia: Its Origin, Affinities and Distribution, Being an Introductory Essay to the Flora of Tasmania*, 1st edn, reprinted from *The Botany of the Antarctic Voyage of HM Discovery Ships* Erebus *and* Terror *in the Years 1839–1843*, Lovell Reeve, London, cxxv.
3 RH Drayton (2000), *Nature's Government: Science, Imperial Britain and the 'Improvement'*

of the World, Yale University Press, New Haven, 129–69.
4 JD Hooker (1899), Reminiscences of Darwin, *Nature*, 60(1547), 187.
5 L Huxley (1918), *Life and Letters of Joseph Dalton Hooker*, John Murray, London, vol 1, 166.
6 Huxley, *Life and Letters of Joseph Dalton Hooker*, 163.
7 JD Hooker to WJ Hooker, 7 September 1840, Letters & Journal 1839–1843, JDH/1/3, Archives, Royal Botanic Gardens, Kew, Richmond.
8 R Gunn to JD Hooker, 8 December 1843, KDC218, 1865–1900, Archives, Royal Botanic Gardens, Kew, Richmond.
9 JD Hooker to R Gunn, 13 May 1844, Gunn correspondence 1833–1854, Mitchell and Dixson Libraries, State Library of New South Wales, Sydney.
10 R Gunn to JD Hooker, 26 September 1844, KDC218, Kew Archives.
11 JD Hooker, (1844), Note on the cider tree, *London Journal of Botany*, iii; WH Breton, (1846), 'Excursion to the western range, Tasmania', *Tasmanian Journal of Natural Science, Agriculture, Statistics &c*, 2(7), 140–41.
12 JD Hooker to R Gunn, October 1844, Gunn correspondence, Mitchell Library.
13 R Gunn to JD Hooker, 28 May 1845, KDC218, Kew Archives (emphasis in original).
14 R Gunn to WJ Hooker, 31 March 1837, in TE Burns & JR Skemp (eds) (1961), *Van Diemen's Land Correspondents: Letters from RC Gunn, RW Lawrence, Jorgen Jorgenson, Sir John Franklin and Others to Sir William J Hooker, 1827–1849*, vol 14, The records of the Queen Victoria Museum, Launceston, Queen Victoria Museum, Launceston, 62.
15 R Gunn to WJ Hooker, 31 March 1837, in *Van Diemen's Land Correspondents*, 62.
16 Hooker, *On the Flora of Australia*, cxxv.
17 R Gunn to JD Hooker, 20 August 1844, KDC218, Kew Archives.
18 Mueller to WJ Hooker, 5 April 1855, in RW Home et al (eds) (1998), *Regardfully Yours: Selected Correspondence of Ferdinand von Mueller 1840–1859*, vol 1, Peter Lang, Bern, 215; PF Stevens (1997), 'JD Hooker, George Bentham, Asa Gray and Ferdinand Mueller on species limits in theory and practice: A mid-19th century debate and its repercussions', *Historical*

Records of Australian Science 11(3), 350. Joachim Steetz (1804–62) was a lecturer at the Hamburg botanic garden; Conrad Gideon Scuchhardt (1829–92) was director of the botanic garden at Waldau Agricultural College near Koningsberg: neither visited Australia (see Home et al, *Regardfully Yours*, 556, 559).

19 R Gunn to JD Hooker, 4 July 1841, JDH/2/10, Royal Botanic Gardens, Kew, Richmond.

20 JD Hooker to R Gunn, October 1844, Gunn correspondence, Mitchell Library.

21 Hooker, *On the Flora of Australia*, xxx.

22 Anon (Maxwell Tylden Masters?) (1873), editorial, *Gardener's Chronicle*, 49 (6 December), 1631.

23 Hooker, *On the Flora of Australia*, cxxv.

24 W Baulch, 'Ronald Campbell Gunn', in *Van Diemen's Land Correspondents*, xv.

John MacGillivray: merits all his own

1 A David (1995), *The Voyage of HMS* Herald *to Australia and the South-West Pacific 1852–1861 under the Command of Captain Henry Mangles Denham*, Melbourne University Press, Melbourne, 76.

2 See JH Calaby (1967), 'MacGillivray, John (1821–1867)', *Australian Dictionary of Biography*, Melbourne University Press, vol 2, 167–68. Calaby states that 'there is a tradition among Australian naturalists that his career "ended on a low rung"'.

3 R Ralph (1993), 'John MacGillivray: His life and work', Archives of Natural History, 20(2), 193.

4 MacGillivray to J Harley (nd), Harley correspondence, Leicester and Rutland Archive, 13 D56/7, f 42.

5 MacGillivray to J Gould, 12–19 May 1849, *Voyage of HMS* Rattlesnake: *Copies of Letters on Natural History of the Voyage 1847–1849*, Mitchell Library, microfilm FM4/2231.

6 MacGillivray to E Forbes, 1 February 1849, *Voyage of HMS* Rattlesnake.

7 R Ralph (1999), *William MacGillivray: Creatures of Air, Land and Sea*, Merrell Holbertson and The Natural History Museum, London, 54.

8 MacGillivray to E Forbes, 1 February 1849, *Voyage of HMS* Rattlesnake.

9 MacGillivray to E Forbes, 1 February 1849, *Voyage of HMS* Rattlesnake.

10 Ralph, *William MacGillivray*, 50.

11 WJ Hooker to F Beaufort, 1 January

1852, National Library of Australia, UK Hydrographic Office records, letters pre-1857, microfilm M2322–2324.

12 E Forbes (1848), 'Letters from J MacGillivray, Esq, naturalist to HM Surveying Ship Rattlesnake, Capt Stanley, RN', *The Annals and Magazine of Natural History* 2, 31.

13 MacGillivray to JD Hooker, 27 May 1848, *Voyage of HMS* Rattlesnake.

14 MacGillivray to E Forbes, 2 April 1849, *Voyage of HMS* Rattlesnake.

15 MacGillivray to White, 22 January 1848, *Voyage of HMS* Rattlesnake.

16 'Report of an enquiry upon Mr MacGillivray, 26 April 1855', [UK] National Maritime Museum, FRE: 205.

17 HM Denham to WJ Hooker, 27 April 1855, Kew Directors' correspondence, vol 35, f 134.

18 TH Huxley to JD Hooker, 17 November 1855, Imperial College, Huxley papers, 2:196.

19 J Gray to TH Huxley, 17 March 1856, Huxley papers, 17:109.

20 G Krefft (1874), 'Autobiographical notes and report', National Library of Australia, MS 3321.

21 'Shipping', *Sydney Morning Herald*, 8 February 1858, 4.

22 MacGillivray, 'Reminiscences of New Caledonia', *Sydney Morning Herald*, 21–30 March 1860.

23 *Sydney Morning Herald*, 4 Jan–1 April 1862.

24 MacGillivray to EP Ramsay, 1 July 1866, Mitchell Library, EP Ramsay papers, 1860–1912, MSS 563/3.

25 MacGillivray to EP Ramsay, 1 March 1867, Ramsay papers.

26 MacGillivray to EP Ramsay, 1 March 1867, Ramsay papers.

27 MacGillivray's petrel, also known as the Fiji petrel was actually collected after MacGillivray's dismissal from the *Herald* by the ship's surgeon, FM Raynor, in October 1855.

28 I am currently undertaking this longer study is the basis of my doctoral thesis at the Australian National University, Canberra.

29 'Mr John MacGillivray', *Clarence and Richmond Rivers Examiner*, 11 June 1867, 2.

Explorers and traders of South Papua

1 J Moresby (1876), *Discoveries and Surveys in New Guinea and the D'Entrecasteaux Islands*, John Murray, London, 197.

2 Moresby, *Discoveries and Surveys*, 201–202.

3 A Pawley (2007), The origins of early Lapita culture: The testimony of historical linguistics, in S Bedford, C Sand & SP Connaughton (eds), *Oceanic Explorations Lapita and Western Pacific Settlement Terra Australis* 26, ANU E Press, Canberra, 27.

4 G Summerhayes and J Allen (2007), Lapita writ small? Revisiting the Austronesian colonisation of the Papuan south coast, in Bedford, Sand & Connaughton (eds), *Oceanic Explorations*, 100.

5 B Malinowski (1922), *Argonauts of the Western Pacific*, Routledge and Kegan Paul, London.

6 PV Kirch (1991), Prehistoric change in western Melanesia, *Annual Review of Anthropology*, 20, 141–65.

7 FR Barton (1910), The annual trading voyage to the Papuan Gulf, in CG Seligman, *The Melanesians of British New Guinea*, Cambridge University Press, Cambridge, 96.

8 Barton, The annual trading voyage to the Papuan Gulf, 104.

9 B Malinowski (1915), The natives of Mailu: Preliminary results of the Robert Mond research work in British New Guinea, *Transactions of the Royal Society of South Australia*, xxxix, 494–706: 628.

10 Discussed in D Lipset (1997), *Mangrove Man: Dialogics of Culture in the Sepik Estuary*, Cambridge University Press, Cambridge, 23.

11 Malinowski, The natives of Mailu, 575.

12 Malinowski, The natives of Mailu, 629.

13 D Battaglia (1990), *On the Bones of the Serpent: Person, Memory, and Mortality in Sabarl Island Society*, University of Chicago Press, Chicago, 119.

Days of desolation on the New Guinea Coast

1 T Huxley to H Heathorn, [no date] Tuesday morning Rattlesnake 1850, Thomas Huxley/ Henrietta Heathorn correspondence 1847–1854, Thomas Henry Huxley papers, Imperial College of Science and Technology, contained in the microform collection, *Darwin, Huxley and the Natural Sciences*, unit 2, mfm 054.

2 T Huxley to H Heathorn, 1 September 1848, Huxley/Heathorn correspondence.

3 J MacGillivray (1852) *Narrative of the Voyage of the HMS* Rattlesnake, T&W Boone, London, 84.

4 This does not include items collected by crew

and officers of the *Bramble*, or those things which were kept as personal collections by officers of the *Rattlesnake*.

5 MacGillivray, *Narrative*, 168–81.

6 M Lepowsky (1993), *Fruit of the Motherland: Gender in an Egalitarian Society*, Columbia University Press, New York, 55.

7 J Huxley (ed) (1936), *TH Huxley's Diary of the Voyage of HMS* Rattlesnake, Doubleday, New York, 148. In posing the question, Huxley prefaced a serious point of anthropological and ethnological debate that was a focus of the discipline well into the twentieth century.

8 MacGillivray, *Narrative*, vol 1, 200.

9 Lepowsky, *Fruit of the Motherland*, 218.

10 MacGillivray, *Narrative*, vol 1, 227.

11 Huxley, *Diary of the* Rattlesnake, 150.

12 Huxley, *Diary of the* Rattlesnake, 149.

13 Huxley, *Diary of the* Rattlesnake, 201–202.

14 Huxley, *Diary of the* Rattlesnake, 165 and 180.

15 It is worth noting that on this part of the New Guinea coast, and also in the Louisiade Archipelago, women had and have a far greater autonomy and leadership within the community than would be understood from the nineteenth century: Lepowsky, *Fruit of the Motherland* and M Demian (2000), Longing for completion: Toward an aesthetics of work in Suau, *Oceania*, 71(2), 94–109.

16 CJ Card, Diaries 1847–1850, Heritage Collections, State Library of Queensland.

17 Huxley, *Diary of the* Rattlesnake, 183.

18 Card, Diaries, 259; also Huxley, *Diary of the* Rattlesnake, 184.

19 T Huxley to H Heathorn, 14 October 1850, Huxley/Heathorn correspondence.

20 JB Jukes (1847), *Narrative of the Surveying Voyage of HMS Fly*, Boone, London.

21 D Moore (1979), *Islanders and Aborigines at Cape York*, Australian Institute of Aboriginal Studies, Canberra, 226.

Relying on the locals

1 AR Wallace (1969), *My Life: A Record of Events and Opinions*, vol 1, Farnborough, Gregg International, 311.

2 AR Wallace (1853), *A Narrative of Travels on the Amazon and Rio Negro: With an Account of the Native Tribes, and Observations on the Climate, Geology, and Natural History of the Amazon Valley*, Reeve and Co, London, 273.

3 Wallace, *A Narrative of Travels*, 275.

4 Wallace, *My Life*, vol 1, 308.

5 A Williams-Ellis (1966), *Darwin's Moon: A Biography of Alfred Russel Wallace*, Blackie, London, 77–79.

6 AR Wallace (1890), *The Malay Archipelago*, Periplus reprint of the 10th edition, Parkstone, 1.

7 Wallace, *My Life*, vol 1, 336.

8 Wallace, *Malay Archipelago*, 19–26.

9 AR Wallace (1870), 'On the law which has regulated the introduction of new species', in *Contributions to the Theory of Natural Selection: A Series of Essays*, Macmillan, London, Michigan University reprint, 1–25.

10 Wallace, *Malay Archipelago*, 54–68, 71–72.

11 Wallace, *My Life*, vol 1, 242–43.

12 Wallace, *Malay Archipelago*, 119.

13 AR Wallace, 'On the natural history of the Aru Islands', *Supplement to Annals and Magazine of Natural History*, 20, (1857), 473–85, (S38:1857): http://www.wku.edu/~smithch/Wallace/SO38.htm.

14 Wallace, *Malay Archipelago*, 120–21.

15 P van Oosterzee (1997), *Where Worlds Collide: The Wallace Line*, Reed Books, Melbourne.

16 AR Wallace to HW Bates, 4 January 1858, in J Marchant (ed) (1916), *Alfred Russel Wallace: Letters and Reminiscences*, Cassell, London, 67.

17 G Daws & M Fujita (1999), *Archipelago: The Islands of Indonesia from the Nineteenth-Century Discoveries of Alfred Russel Wallace to the Fate of Forests and Reefs in the Twenty-First Century*, University of California Press, Berkeley, 79.

18 Wallace, *Malay Archipelago*, 309. In Classical geography, Ultima Thule was the most distant or last land known – variously identified as Britain, Ireland or Iceland.

19 Wallace, *Malay Archipelago*, 309.

20 Wallace, *Malay Archipelago*, 312.

21 Wallace, *Malay Archipelago*, 314.

22 Wallace, *Malay Archipelago*, 310–312.

23 Wallace, *Malay Archipelago*, 310–311.

24 Wallace, *Malay Archipelago*, 313–314.

25 Wallace, *Malay Archipelago*, 317–318.

26 Wallace, *Malay Archipelago*, 323–325.

27 Wallace, *Malay Archipelago*, 326.

28 Wallace, *Malay Archipelago*, 328–29.

29 Wallace, *Malay Archipelago*, 337–39.

30 Wallace, *Malay Archipelago*, 336.

31 Wallace, *Malay Archipelago*, 369.

Instruments and expeditions

1 J Browne (1995), *Charles Darwin: Voyaging*, Jonathan Cape, London, 178.

2 R FitzRoy (1839), *Narrative of the Surveying Voyages of His Majesty's Ships* Adventure *and* Beagle *between the Years 1826 and 1836*, Henry Colburn, London, vol 2, 18.

3 FitzRoy, *Narrative*, 330.

4 FitzRoy, *Narrative*, 396.

5 FitzRoy, *Narrative*, 343.

6 FitzRoy, *Narratives*, 91n.

7 FitzRoy, *Narrative*, appendix, 62.

8 FitzRoy, *Narrative*, 245.

9 F Burkhardt & S Smith (eds) (1985), *The Correspondence of Charles Darwin 1821–1836*, Cambridge University Press, Cambridge, vol 1, 149.

10 Janet Browne (*Charles Darwin*, 171, n 1) states that Darwin bought his own barometer. The *Correspondence* (note 9, vol 1, page 175, n 1) asserts that Darwin also took with him a set of aneroid barometers. This could not be so, as aneroids did not become commercially available until the late 1840s. There is a mountain barometer by John Newman of 122 Regent Street, London, at Down House but it is not clear when this was purchased.

11 FitzRoy, *Narrative*, vol 2, appendix, 30.

12 FitzRoy, *Narrative*, 301.

13 FitzRoy, *Narrative*, 54.

14 CW Thomson (1895), *Report of the Scientific Results of the Voyage of HMS* Challenger *during the Years 1873–76*, Her Majesty's Stationery Office, London, 1(1): 83.

15 *Nature*, 12, 1 July 1875, 173.

16 Major Sabine, 'Magnetic observations', in Fitzroy, *Narrative*, vol 1, 497.

17 FitzRoy, *Narrative*, 24.

18 FitzRoy, *Narrative*, appendix, 329.

19 FitzRoy, *Narrative*, 330.

20 FitzRoy, *Narrative*, 345.

21 *The Correspondence of Charles Darwin*, 1: 143.

22 WAS Burnett (1992), 'Darwin's microscopes', *Microscopy*, 36, 604–27.

23 Thomson, *Report* 1(1), 3–4.

24 Thomson, *Report*, 7

index